Job #: 85566

Author name: Cory

Title of book: Business Ethics

ISBN number: 0387230408

BUSINESS ETHICS
The Ethical Revolution of Minority Shareholders

BUSINESS ETHICS
The Ethical Revolution of
Minority Shareholders

by

JACQUES CORY

 Springer

Jacques Cory
International Business Programs
2 Costa Rica Street
34981 Haifa
Israel
coryj@zahav.net.il

Library of Congress Cataloging-in-Publication Data

A C.I.P. Catalogue record for this book is available
from the Library of Congress.

Cory, Jacques.
    Business ethics: the ethical revolution of minority shareholders
            Includes index.

ISBN 0-7923-7300-6 (hardcover)
© 2001 by Kluwer Academic Publishers

First Springer Science+Business Media, Inc. softcover printing, 2005
ISBN 0-387-23040-8   e-ISBN 0-387- 22914-0   Printed on acid-free paper.

Printed in the United States of America.

9 8 7 6 5 4 3 2 1   SPIN 11319436

springeronline.com

CONTENTS

ACKNOWLEDGEMENTS

The subject of safeguarding the rights of minority shareholders has not received much attention until now. After experiencing more and more cases of wrongdoing throughout my business career, I have decided to explore thoroughly the subject and devote several years to study the matter in a theoretical and empirical research. My approach is mainly ethical, and I do not hesitate to state vehemently my attitude towards the deterioration in business ethics that is mixed with confidence in the new vehicles for the safeguarding of minority shareholders, mainly the Internet, Transparency, Activist Associations and Ethical Funds.

In view of the pertinence of the book's subject, many friends and colleagues have assisted me with their advice and support, and I would like to mention especially Meir Tamari, James Weber, Samuel Holtzman, Jean-Philippe Deschamps, Edvard Kimman, Andrew Pendleton, Anke Martini, Dove Izraeli, Ishak Saporta, Francis Desforges, Nathalie Mourgues, Gad Proper, Rafi Koriat, Daniel Weihs, Peretz Lavie, Rachel Zeiler, Naomi and Itshak Zoran, Dalia Golan, Tali Aharoni, Dietmar Fuchs, Nira Cory and Pierre Kletz.

I would like to express my gratitude to Colette Neuville, President of ADAM, who succeeds almost alone in conducting a crusade against French companies that wrong the rights of minority shareholders. Her courage and perseverance are admirable and her comments were very valuable to me.

I also want to thank my excellent editor Allard Winterink, who has made a remarkable contribution to my book, had confidence in me, and assisted in his warm and kind manner to make the project succeed. Special thanks also to Carolyn O'Neil and Jill Garbi for their valuable assistance in editing the book.

Last but not least, my greatest thanks are to my wife Ruthy, and my children, Joseph, Amir and Shirly, who encouraged me throughout the long years of struggle and efforts in order to achieve the expected results, with many personal sacrifices, but with an unequalled cohesion.

The frank and open approach and the conclusions and recommendations on the touchy issues of this book may not be pleasing to some of the readers. I hope that even more readers will react favorably and even with enthusiasm to the subjects treated. My attitude is simple, I would like to shake the indifference on the cases of wrongdoing to minority shareholders in order to bring the subject to the forefront of the public's interest. If this book will contribute even slightly to this purpose, I will have attained my goal.

ABOUT THE AUTHOR

Jacques Cory is a businessman with a background in Economics and Business Administration (MBA from Insead, France), who has encountered many cases of despoiling minority shareholders in his long international career in top-level positions in the Israeli high-tech industry and in mergers and acquisitions.

Cory decided to write a thesis and a book on this subject in a frank and open approach, in order to bring the subject to the forefront of the public's interest. He is a member of Transparency International Israel and The Society for Business Ethics, and is very active in the Israeli business ethics community.

Cory is the author of the ethical novel "Beware of Greeks' Presents", to be published in the forthcoming months in Israel by Bimat Kedem Publishers.

1
INTRODUCTION

"Morten: And what are we going to do, when you have made liberal-minded and high-minded men of us?
Dr. Stockman: Then you shall drive all the wolves out of the country, my boys!"
(Ibsen, An Enemy of the People, Act V)

The theoretical and empirical research of this book describes how the traditional safeguards of the rights of minority shareholders have failed in their duty and how those shareholders have remained practically without any protection against the arbitrariness of the companies and majority shareholders. The law, the SEC, society, boards of directors, independent directors, auditors, analysts, underwriters and the press have remained in many cases worthless panaceas. Nevertheless, in the Ethics of 2000 new vehicles have been developed for the protection of minority shareholders, mainly the Internet, transparency, activist associations and ethical funds. Those vehicles give the shareholders at least the chance to understand the pattern and methods that are utilized to wrong them and give them a viable alternative for investment in ethical funds.

The new vehicles will prevent minority shareholders from using the Armageddon weapon, by ceasing to invest in the stock exchange and causing the collapse of the system, that discriminates against them. The preconditions for the ethical revolution of minority shareholders do exist, but they are insufficient as other conditions are needed to be met, such as the ostracizing of unethical managers by society, appointment of ethical CEOs to head the companies, and above all - giving an equal weight to financial and operational performance (the hardware), as well as to ethics and integrity (the software).

The book is based on qualitative and inductive research. All the cases presented in it are based on current events and try to find the common aspects and basic rules that govern wrongdoing to minority shareholders. In the four cases of US, French and Israeli companies, most of them in high-tech, the minority shareholders lose almost all their investment. Those are not exceptional cases but rather the norm in many companies, which is illustrated by qualitative cases, without being able of course to quantify them. Case studies are the preferred strategy when 'how' and 'why' questions are being posed. The purpose of this book is therefore to analyze why and how companies do not act ethically toward their minority shareholders, not how many, not which, not to what degree and not where.

The evolution of business ethics in the last ten years of the twentieth century has been accomplished in parallel to the political, social and economical world developments. It is not a coincidence that in the decade where the Iron Curtain has collapsed, most of the world conflicts have been resolved, and the western world has accomplished unprecedented economic achievements, business ethics has started to become an inevitable norm in most of the developed world, especially in the United States.

The notions of quality, ecology and service have become predominant, employee harassment has become illegal, companies contribute more and more to the community, and ethics has ceased to be an oxymoron. Only in one field has there been practically no progress – ethics in the relations between companies and minority shareholders. The officers of the companies pledge allegiance uniquely to the majority shareholders, or rather to the shareholders who control the boards of directors of their companies, even if they hold only the minority of the shares. Those shareholders will be referred to in this book as majority shareholders.

The most advanced country in its protection toward minority shareholders is probably the United States. Following the scandals of the `80s, the SEC, public opinion, the press and the academic world are rather sensitive to this subject, although the other ethical norms are much more applied. France and Israel have done very little to safeguard the rights of minority shareholders, and it is high time that the public and academic world will put this topic in the forefront of the business world's attention.

The goal is not to promote altruism in business, although it is a valuable cause. It is evident that if majority shareholders, together with the companies' management, will continue to wrong the interests of minority shareholders, the latter will cease to invest in the stock exchange and cause the inevitable and irreversible collapse of the value of shares, the impossibility to raise funds in the stock exchange, and finally – a much worse recession than in the `30s.

Minority shareholders can almost never count on a legal defense as they have a very slight chance against the companies, which are armed with powerful legal defense and public relations teams as well as 'unlimited' time and resources to devote to winning their cases. In most cases, minority shareholders are individuals with limited resources who do not have the time, funds and know-how to fight against large companies. The press sympathizes in many cases with the companies, particularly those that hold shares in the newspapers, provide advertising budgets or are closely tied with the press in other businesses.

It is true that in many cases minority shareholders are large financial institutions, funds that manage pensions and savings of employees, investment funds and others; that have much more power than the individual shareholders. In those cases, majority shareholders tend to compensate partially the funds, but only in cases where the funds' managers use their strength, threaten or actually sue the companies. In many cases, the funds' management does not learn at all that they have been wronged, as they manage thousands of investments, or they do not want to devote the management attention and the money to recuperate the sums that are only a small fraction of their total investment. In other cases, the interests of the funds differ from the interests of the small shareholders that invest their savings in the funds, and they sacrifice those interests on behalf of other interests that they have in their agenda.

One reason for the 'clean' conscience of the managers of the companies, that despoil the rights of the minority shareholders is the lack of personification of those shareholders, who are in most cases small shareholders who do not know anybody in the companies, and who are usually interested in obtaining prompt profits. They are perceived by the managers and majority shareholders as speculators who cannot cause them any harm. It is much easier to commit a wrongdoing toward somebody that you do not know and do not appreciate, especially if you are convinced that you are right. However, the managers have a personal interest in their companies, and they perceive their missions beyond the immediate profits, pledging allegiance to the majority shareholders that have often founded the companies, possess the control and can remunerate and fire them.

Small minority shareholders do not know in many cases that they have been wronged, as they do not have links between them, and the schemes are performed in the shade, far from the public eye. Transparency is therefore necessary in order to dissipate the fog that the wrongdoers want to prevail. The press often sympathizes with the large companies and rarely agrees to divulge scandalous cases. Other businessmen who might have lost large sums as minority shareholders do not complain against their colleagues due to the law of Omerta, which states that you do not file complaints against a colleague. Most businessmen prefer not to open Pandora's boxes often containing similar cases of their doing. In extreme cases, you can always buy the silence of your colleagues or compensate them in an indirect way.

The minority shareholder can always complain to the Securities Exchange Commission (SEC) in the US, the Commission des Operations de Bourse (COB) in France, the Israeli Securities Commission, or similar organizations, that will be referred hereinafter in this book as 'SEC'. But these organizations are governmental, with limited budgets, and have an agenda that does not always include safeguarding the interests of minority shareholders. Those are not as politically influential as the large companies that contribute funds to the political parties who govern the countries and who name the heads of the SEC. Even worse, the heads of the government organizations receive top-level positions in the companies they controlled, after they leave their organizations. The least that could be done is that those officials will not be allowed to work in business organizations and will get a lifetime pension, like judges, in order to ensure their objectivity.

How is it possible to implement this change of attitude? One has to start probably at the top of the companies, the CEOs, as it is they who ultimately determine the ethical climate of their companies. Unfortunately, the companies are the last vestige of the dictatorial regimes. Most of the world, especially after the collapse of the USSR, is ruled by democracies. All western countries have laws on equal rights regardless of gender, race and religion. Only one domain of human activity remains dictatorial – the companies, where the CEOs and the majority shareholders have absolute power and rule the companies.

The greatest danger for minority shareholders consists in the holy alliance between the executives of the companies and the majority shareholders, who appoint and remunerate them. Those executives involve themselves in the quarry, by receiving shares and warrants of the companies in very advantageous terms that enable them to get rich with almost no risks.

One of the most serious problems is that the shareholders who control the companies almost always have insider information. In some countries it is legal to benefit from such information in the buying and selling of shares in the stock exchange, and in others it is very difficult to prevent the use of such information due to the collaboration of the CEOs. In the bullish periods, majority shareholders often succeed in selling a sufficient amount of shares in the stock exchange or at public offerings, with very high price to earnings ratios, recouping their initial investment.

From this moment on they no longer risk their money, and even if the company gets into trouble, the only shareholders who lose their investment are minority shareholders who were the last to purchase the shares at the high prices. Therefore, majority shareholders tend to speculate very often, much more than if they would have risked their own money; as they know in advance, much ahead of the public and minority shareholders, when it is worthwhile to sell their shares - if the condition of the company will deteriorate, and when it is preferable to buy shares - if the financial conditions of the company are expected to improve.

The executives and majority shareholders who commit unethical and unlawful acts are not ostracized by society. On the contrary, very often, they are admired and envied by their colleagues who would have behaved similarly if they only had the opportunity. They are treated as 'smart guys' who take advantage of the good opportunities that they encounter. Man is before everything a social animal and it is imperative that businessmen who are unethical will be treated as outcasts, banned by society and despised by their fellows.

In recent years a revolution has occurred in the publication of data on the Internet. Most of the quoted companies have a site on the Internet and stock talk groups comprised mainly of minority shareholders, where information and misinformation is shared between the shareholders who have access to the Internet. It is pure democracy, as in the agora of Athens, where all citizens had the right to participate. Information about future wrongdoing to minority shareholders can be divulged in advance and one has only to read it and sell his shares, while there is still time.

The full transparency of companies, via the Internet and ethical reports, could safeguard ethics, even if it is achieved through the assistance of whistle-blowers. Transparency compels every employee to adopt an ethical conduct, as his conduct could be published on the Internet and the press or scrutinized by activist associations, so that his family, friends and congregation would learn of his conduct.

The implementation of ethics is assisted by the ethical funds. These funds were established primarily in the United States, but are also very influential in Canada, the Netherlands and Great Britain in the last ten years. They comprise investments of more than two trillion US$ in the United States and have succeeded in obtaining financial results above the average of the US stock exchange, while keeping very strict ethical screening. The minority shareholders will have to collaborate with those funds and buy only shares of ethical companies.

Ethical investing is screened to reflect ethical, environmental, social, political, or moral values. It examines the social records of companies in local community affairs, labor, minority and gender relations, military and nuclear production, product quality, approach to customers, suppliers and shareholders, and avoidance of sales of tobacco, alcohol, pornography or gambling products.

In the last decade of the 20th century we witnessed in the US and in France, but not in Israel, an effervescence in social and other activism of shareholders, and in many cases they have succeeded in changing the initiatives of very large companies, especially in the United States. One of the main activist associations that has launched a mission to fight corruption is Transparency International, which published in 1999 a survey on the level of world corruption, ranking the United States in 18th place, Israel in 20th and France in 22nd. Less corrupted countries are Denmark in 1st place, Canada - 5th, the Netherlands - 8th, UK - 13th and Germany - 14th place.

An example of an activist association uniquely dedicated to safeguarding minority shareholders is l'Association pour la Defense des Actionnaires Minoritaires (ADAM), which was founded in 1991 in France by its president Colette Neuville. It has conducted very important campaigns to safeguard and prevent wrongdoing to the interests of French minority shareholders, as described in one of the cases of this book. The ethical struggle of minority shareholders conducted by activist associations has to be fought vigorously but ethically, among others because the majority shareholders will always be the strongest while using unethical methods, and the minority shareholders will lose the legitimacy of their campaign.

But it is not enough that the campaign of minority shareholders will be fought vigorously, ethically and courageously. It has to be fought also cleverly and even cunningly. Ulysses did not succeed to win the Trojans until he introduced the Trojan Horse. In our case, the Trojan Horses in the long Odyssey of the minority shareholders are mainly the Internet and Transparency. Those vehicles seem very innocent but have a tremendous power that will benefit the weaker side of this campaign.

The majority shareholders, the boards of directors, and the executives of the unethical companies who work for them tend to prefer an obscure and opaque environment for their activities. Even when they have to disclose their intentions in prospectuses, press releases or financial statements they do it in many cases in such a way that average readers are confused with the facts, do not perceive the double meaning of their terminology, and cannot read and understand all the material that is handed to them. This book will show how Transparency and Internet give the minority shareholders and stakeholders the opportunity to reach the truth, in the most efficient, clear and precise way.

The book concludes that if the new vehicles will be preponderant, and the other conditions will be met, then - led by activist associations - minority shareholders will be properly organized, motivated and conscious of their strength, thereby enabling them to win their fight and safeguard their interests while behaving ethically. After all, the majority shareholders will never give up their privileges willingly. The victory of this revolution will be the victory of all of us, as everybody nowadays is considered to be a minority shareholder and a stakeholder of a company, whether directly or indirectly through our pension funds, or as a client, supplier, subcontractor, employee, member of a community and citizen of a country in which the companies operate.

2
THE INEFFICIENT SAFEGUARDS OF THE MINORITY SHAREHOLDERS

"Selon que vous serez puissant ou miserable,
Les jugements de Cour vous rendront blanc ou noir."
According to your mighty or miserable position,
The judgment of court will render you white or black.
(La Fontaine, Fables, Livre septieme, Fable I)

There has been no major improvement, unfortunately, since the times of La Fontaine's fables until today. While presuming that the judges are incorruptible, we have to admit that individual, weak, minority shareholders, who do not have the time, means, and the assistance of the best lawyers, do not have much opportunity to win a case against the tycoons of finance. In paraphrasing the title of the film 'The Untouchables', which tells the story of how Al Capone was sent to jail by untouchable government agents, who could not be corrupted, we notice how the norms have evolved nowadays and how the large companies are now untouchables, as the minority shareholders cannot touch them or undermine their power if they have to confront them in court.

The purpose of this book is to render the unethical businessmen 'untouchables' in the religious sense of the word, like the caste in India, so that nobody would approach them, associate with them, or pay any attention to them. This attitude would be in contradiction to the present veneration that they enjoy from many of their colleagues. The unethical businessmen will be ostracized and apprehended by their Achilles' heel, which is the importance that they give to their image in society. Their donations will be refused by universities. They will receive no more honorary doctorates or legion of honor. Impossible to imprison them due to their power, they should be treated socially as Mafia outcasts.

All that is legal is not necessarily ethical, and all that is unethical is not necessarily illegal. It could be legal to pour toxic materials into a river, but this is certainly unethical and harmful. Laws can change, but ethics is much more immutable. "Even more, laws themselves must be governed by moral criteria, which gives rise to the classic distinction between just and unjust laws. Thus, a law that violates a person's dignity (sanctioning slavery, for example) is not just and therefore cannot be accepted and observed... a just law... must be observed, not for merely practical reasons (to avoid punishment, for example) but also for moral reasons: there is an ethical obligation to observe it." (Harvey, Business Ethics, A European Approach, Argandona, Business, law and regulation: ethical issues, p. 128-129) We should educate people to behave ethically exactly as we educate them to obey the laws. Aristotle has said that in order to know how to conduct we have to observe a just person. This maxim is somehow difficult to observe in the modern business world, but we can nevertheless compare ourselves to businessmen, who are relatively ethical.

"Ethics is above law and is also the source of the power of the law to oblige morally. Laws are not something sacred, as Latin culture sometimes pretends: they are no more (and no less) than an instrument at the service of the common good of society. They are not an obstacle that must be knocked down, jumped over or bypassed. They should be respected as a condition for the proper functioning of society, and even as a condition for personal freedom. (This notwithstanding, it must be recognized that in practice many laws may be defective or even immoral, and therefore not compelling.)" (Harvey, Business Ethics, A European Approach, Argandona, Business, law and regulation: ethical issues, p.130) This is the reason why in the polemic between legality and ethics in business, the ethical considerations should be predominant, because ethics is above the law, it is almost universal and immutable, while laws are conjunctural, national and often unjust.

One of the most acute dilemmas of managers is the dilemma between cases, which a priori seem equally ethical, but from different angles. Not the dilemmas between just and unjust situations, as in this case the choice is obvious, although it is not so simple for many businessmen. But the dilemma between two just positions is much more intricate, as it is incrusted in our basic values. "Four such dilemmas are so common to our experience that they stand as models, patterns, or paradigms. They are: Truth versus loyalty, Individual versus community. Short-term versus long-term. Justice versus mercy." (Kidder, How Good People Make Tough Choices, p.18) Kidder and many other authors on ethics prefer ultimately truth to loyalty, as it is better to divulge cases that are not ethical than to remain loyal toward a management that is not ethical.

The author gives examples of loyalty toward Hitler, Mao, Stalin, Sadam Hussein, or even Richard Nixon, which caused great damages to humanity, but we should also mention the fate of those who preferred truth over loyalty and who ended up in suffering atrociously. Between the individual and community he prefers community, although he mentions that if he was a Soviet citizen he would perhaps prefer the individual. Between short-term and long-term he prefers long-term, as we see how the financial scandals of the `80s, which were based on immediate gains in the short-term, were detrimental to society. And if he would have to choose between justice and mercy, Kidder would have opted for mercy, which signifies for him compassion and love. As he can imagine a world so full of love that there would be no need for justice, but he cannot imagine a world so full of justice that we would not need any more love.

One of the most well-known cases that illustrates those conflicts is the controversial case of Shylock, the Jew of Venice, who insisted on preferring justice over mercy, by getting the pound of flesh that he asked for as a collateral. This is the case of an individual who feels persecuted by the community and wants to avenge himself. This is the case of a person who knows that if he is satisfied in the short term he is going to lose in the long term. This is the case of the businessman who has his own truth, which is opposed to the loyalty that he owes to the Duke of Venice. And Shylock exposes his point of view in the well-known dialogue with Salarino:

"I am a Jew. Hath not a Jew eyes? hath not a Jew hands, organs, dimensions, senses, affections, passions? fed with the same food, hurt with the same weapons, subject to the same diseases, healed by the same means, warmed and cooled by the same winter and summer, as a Christian is? If you prick us, do we not bleed? if you tickle us do we not laugh? if you poison us, do we not die? and if you wrong us, shall we not revenge?"
(Shakespeare, The Merchant of Venice, Act III, Scene I, p. 203-4)

Shakespeare unties the drama in a manner that favors ethics as being stronger than law, morals being stronger than a given promise. But Shakespeare's ethics is quite equivocal, as it is applied against a Jew, who is treated by the Duke as a stranger. Would the same ethics be implemented if the situation was opposite, and Shylock was a poor Jew who owed money to Antonio, the Merchant of Venice, a Christian originating from an ancient Venetian family? Would we ask Antonio to conduct himself ethically toward a poor Jew in order to prove Christian mercy toward him? The issue of double standards is emphasized here in the most acerbic manner, because in order to conduct ourselves ethically we should apply our ethics first of all toward the weak, the poor, the strangers, and in the cases of this book toward the minority shareholders, who do not have in most cases the possibility to confront the mighty majority shareholders in court.

True ethics is revealed only when you do not have a sympathizing Duke of Venice and a collaborating population on your side... Clemency toward the mighty at the expense of the weak is the height of hypocrisy, and unfortunately this is what is practiced in many cases where the mighty and rich are brought to justice. If a poor thief steals a few hundred dollars he is sentenced to jail for many years, but if an Israeli financial tycoon is found guilty of manipulating the price of the shares of his bank, causing the Israeli minority shareholders and the state of Israel billions of dollars in losses, he is not even sent to jail.

However, we should inlay in golden characters the speech of Portia, who appears at the court disguised as a jurist doctor, and hang it on the walls of all the board rooms in modern companies to be applied for stakeholders and minority shareholders.
"But mercy is above this sceptred sway,
It is enthroned in the hearts of kings,
It is an attribute to God himself,
And earthly power doth then show likest God's
When mercy seasons justice. Therefore, Jew,
Though justice be thy plea, consider this,
That in the course of justice none of us
Should see salvation: we do pray for mercy,
And that same prayer doth teach us all to render
The deeds of mercy."
(Shakespeare, The Merchant of Venice, Act IV, Scene I, p. 211)

The ancient maxim, which says 'if it ain't illegal, it must be ethical', is completely erroneous, as the difference between ethics and law is as the difference between the enforceable and the unenforceable. "Law is a kind of condensation of ethics into codification: It reflects areas of moral agreement so broad that the society comes together and says, 'This ethical behavior shall be mandated.' But Moulton's distinctions also make something else clear: When ethics collapse, the law rushes in to fill the void. Why? Because regulation is essential to sustain any kind of human experience involving two or more people. The choice is not, 'Will society be regulated or unregulated?' The choice is only between unenforceable self-regulation and enforceable legal regulation... Surely a powerful indicator of ethical decay is the glut of new laws – and new lawyers – spilling onto the market each year." (Kidder, How Good People Make Tough Choices, p.68-69)

History is full of examples of how kingdoms, which were lacking ethics, have collapsed, and how regimes that were governed by so-called very humane laws and an exemplary constitution which were not implemented, as in the case of the Soviet Union, have also collapsed. The economic anarchy which prevailed in Italy in the `80s is another example of how the lack of obedience to the law, or even more to ethics, could be harmful to the economic progress.

Should we obey immoral laws? The Nuremberg tribunal has categorically decided – no! But where is the limit between disobedience and anarchy? The English, who judged at those trials, were confronting at the same time the disobedience to the laws of the British Empire from the same Jews who were the victims of the Nazis and wanted to emigrate to Israel. The British arrested thousands of illegal immigrants who returned to their homeland after having survived the Holocaust, and sent them back to Europe or imprisoned them in concentration camps in Cyprus until 1948. The Americans had racist laws enforced until the '70s and only the Civil Rights Movement, headed by Martin Luther King, succeeded in shaking the American conscience and changing the laws and the implementation of the laws.

The companies are ready to invest considerable amounts in trials, which are much larger than the damages they would have to pay to the minority shareholders or the government institutions. GE preferred to pay $30 million in direct and indirect costs during a trial in which the government sued them for the amount of $10 million in damages for price fixing. Ultimately, the company was acquitted, and those who most benefited from the trial were the lawyers, while the shareholders, the government and other stakeholders lost. And this is the case of a trial against the American government. How can we ask from a poor individual shareholder to finance such astronomical sums, while the company will opt almost always to prefer the trial where it feels strong in comparison to the shareholders? We will analyze later on in the empirical part the class-actions and see how, effectively, it is almost impossible for a shareholder to win a case against the mighty companies.

According to Monks, the decision of companies to obey or disobey the law is simply a profit and loss decision. The company checks if the cost of infringement of the law actualized by the probability to be discovered, brought to justice, and punished (there is almost no risk to be imprisoned), is equal to the cost of obedience to the law. If the cost is inferior, the company will prefer to infringe the law. This is why it is imperative that at the head of each company should stand an ethical CEO, with impeccable integrity and ethics, who will not just calculate impersonal feasibility studies on the benefits of obeying the law. We could try to make audits on the adherence to laws, augment the damages paid by companies, and so on, but the companies, with their infinite funds, their masses of lawyers and experts, and their immeasurable patience will win almost inevitably in court against the government, the stakeholders and the minority shareholders. They feel themselves stronger than all those organizations and individuals, and the only way to beat them is to change their attitude de profundis.

The Jewish religion teaches us that a just person builds a fence around the law, as the ethical man has to observe the ethical norms, which are much wider than the law. On the other hand, the modern lawyers seek loopholes in the law and try to reduce the implementation of the law to a minimum, which is in complete contradiction to Jewish law. It is therefore, practically impossible to rely only on the law, which many influential companies and lawyers try to reduce to a minimum, and we have to adhere to the ethical rules which are much wider than the law.

An extremely important aspect, which prevents the minority shareholders in most of the cases to resort to the law, is the time elapsing between the wrongdoing and the decision of the court. Besides the resources that the shareholder lacks, the risk that he incurs, and the loss of health, this excessively long time makes a trial almost prohibitive. In 1990 Kuwait was invaded by Iraq. The country was looted, thousands of citizens were murdered or mistreated, many others fled the country. A country that was once one of the richest in the world was completely ruined. The United States, which decided to intervene, did so only six months after the invasion, while it was practically too late. We say that time is of the essence, and time is an essential factor in international relations as it is also with the rights of minority shareholders. Even if the law can assist ultimately the minority shareholders, if it occurs many years after they lost their money, it is too late to remedy effectively the wrongdoing.

Of all the maxims that differentiate law from ethics, the most salient is probably caveat emptor, which means that the buyer should always beware. Everything is therefore permitted to the seller if it is legal, and it is the buyer of the product or of the stock who should beware not to be wronged. The author of this book maintains that if it is impossible to rely upon the ethics of the seller, it is preferable to abstain from buying the product or the stock, even if it is a bargain, as it is preferable to pay a higher price to an ethical seller than a lower price to an unethical one. The reason is that if you have to beware of the quality, the delivery, the service and so on, the effective price of the unethical seller is much higher than the effective price of the ethical seller.

Nevertheless, there is some evolution in this respect, and the tendency today in many cases is to make the seller beware and advise the buyer of potential defects of the products. This occurs mainly if there is a law requiring it like in the pharmaceutical industry or in the case of McPherson v. Buick in 1916. But do we need to disclose everything to the public? "We need to ask, 'Why in the case of physicians and therapists, as well as for other professionals such as attorneys, clergy, and journalists, is confidentiality so well protected in the law?'.... The duty to warn is limited in these relationships precisely because it is important to protect privacy and fairness, on the one hand, and encourage people to utilize professional help, on the other hand. Thus society forgoes certain benefits that might be derived from disclosure in order to protect other interests." (May, Business Ethics and the Law, p. 19-20)

Ethical thinking and character bring about the ethical conduct, which is different from legal conduct, as the law defines what is permitted and prohibited, while ethics defines what should be done. If the law in the 21st century will be driven by ethics as maintained by certain specialists, it is needed to make a thorough reform in the legal system, in France in particular, as it permits in many cases, especially in the commerce courts – tribunal de commerce, to transgress the rights of the minority shareholders as will be examined in the case of the French company.

The campaigns against arbitrary decisions of the commerce courts conducted by such important persons such as Mme. Neuville, President of ADAM, the association for the protection of the minority shareholders, will undoubtedly have a positive result. This reform will probably not assist the minority shareholders of the French company, which were wronged in one of the cases of this book by their company and were fined with hundreds of thousands francs by the commerce court, but it will assist the minority shareholders of the year 2000 and beyond. Until then, the shareholders could still resort to the appeal court, Cour d'appel and then the supreme court, Cour de cassation, a procedure which is nevertheless very long and costly.

Monks describes in his outstanding book 'The Emperor's Nightingale' the seven panaceas that are supposed to safeguard the corporate accountability. Those panaceas are really not effective cures, although they give a false sense of comfort that is more dangerous than the total lack of cure. The first panacea is the CEO philosopher-king, who is supposed to distribute evenly the goods of the company amongst the stakeholders. Unfortunately, the CEOs today exercise near-monarchic power, and they are free to advance their own personal interests in compensation, even to the point of harming the interests of shareholders. "Institutional Shareholder Services (ISS) found that, in 1992, the top 15 individuals in each company received 97 percent of the stock options issued to all employees. Business Week wrote for all to read that 'the 200 largest corporations set aside nearly 10 percent of their stock for top executives', adding that 'in almost all cases, moreover, it's the superstar CEO who takes the lion's share of these stock rewards." (Monks, The Emperor's Nightingale, p.62) The second panacea says that if a state and/or federal charter sets proper limits, then the corporation can serve the common good. This chart is effectively very weak and is practically non-existing in multinationals.

The third panacea is the independent directors. Those directors are nominated by independent committees and are elected by the shareholders, but in most cases they are effectively appointed by the CEOs of the companies. "Yet true independence – as well as true nominations and elections – remains elusive. How can an individual selected for a well-paying and prestigious job, notwithstanding his or her compliance with the most exhaustive legal criteria of 'independence', be expected to stand in judgment of those who accorded him this favor in the interest of an amorphous group of owners? Only men and women of the highest character can do this, but the best solutions cannot depend on character alone... Directors are not 'nominated', they are selected by the incumbent directors (however independent) and the chief executive officer. Shareholders do not 'vote', whether or not they mark a slate card; only those named on the company proxy will be elected. Ultimately, independence is a matter of personal character... the search of such a director requires that we be modern-day Diogenes, lamp in hand. This is not acceptable. We cannot have a system that depends on the luck of stumbling across an occasional honest man." (Monks, The Emperor's Nightingale, p.53-4)

The fourth panacea is the board of directors, well-structured boards, that rank high as a favored solution to governance problems. Monk believes that even corporations with perfectly independent directors and perfectly structured boards can remain insensitive to the needs of the public. The fifth panacea is independent experts. "The experience with 'experts', however is disheartening. The tendency to generate opinions satisfactory to present and prospective customers is strong. 'Fairness' opinions – whether of the prospective value of Time Warner stock, or in the leveraged buyouts that were the source of the Kluge, Heyman, and many other fortunes – have turned out to be wrong, not by percentages but by orders of magnitude." (Monks, The Emperor's Nightingale, p.55)

The sixth panacea is the free press. The most acute problem of this panacea is the large percentage of the press' revenues that derive from advertising, which may impair the impartiality of the press in regard to companies that finance huge advertising budgets. Furthermore, Westinghouse has recently acquired CBS, Disney owns ABC, GE owns NBC, Time Warner owns Fortune and McGraw-Hill owns Business Week. The situation is similar in France and Israel. It is true that there is no protocol of the sages of the media, but it is difficult to expect critics on an unethical company from a newspaper which is owned by a public company and which can be subjected to retaliation in the future with juicy stories on the owners of the newspaper, written by another newspaper which is owned by a competitor company.

The seventh panacea is multiple external constraints, such as the economic constraints of competition and law, the impact of the tax and regulatory schemes, and the constraining influence of social values on corporate decision making. Adam Smith has recommended to rely on the invisible hand that will arrange everything. It is the same blessed hand that brought the worst recession ever in 1929, all the economic crises, stock exchange scandals, inefficiencies in the legal and governmental system, the reliance on the SEC that will solve everything and so on. All those 'cures' are only panaceas, which cannot cure the wrongdoing to minority shareholders. The empirical research of this book will prove in the case studies how all these panaceas without exception proved to be inadequate at the moment of truth. Only new organisms can cure the illnesses of the existing system, as all the other cures have proved to be in most cases worthless panaceas for safeguarding the interests of minority shareholders.

Zola describes in a magnificent way the panacea of the board of directors in his famous book 'L'Argent', The Money. One would think that Zola had participated in hundreds of board meetings in recent days in the US, Israel or France. Only a genius writer like Zola can remain immortal and stay modern, even after more than 100 years. "Saccard avait acheve de mettre la main sur tous les membres du conseil, en les achetant simplement, pour la plupart. Grace a lui, le marquis de Bohain, compromis dans une histoire de pot-de-vin frisant l'escroquerie, pris la main au fond du sac, avait pu etouffer le scandale, en desinteressant la compagnie volee; et il etait devenu ainsi son humble creature, sans cesser de porter haut la tete, fleur de noblesse, le plus bel ornement du conseil. Huret, de meme, depuis que Rougon l'avait chasse, apres le vol de la depeche annoncant la cession de la Venetie, s'etait donne tout entier a la fortune de l'Universelle, la representant au Corps legislatif, pechant pour elle dans les eaux fangeuses de la politique, gardant la plus grosse part de ses effrontes maquignonnages, qui pouvaient, un beau matin, le jeter a Mazas.

Et le vicomte de Robin-Chagot, le vice-president, touchait cent mille francs de prime secrete pour donner sans examen les signatures, pendant les longues absences d'Hamelin; et le banquier Kolb se faisait egalement payer sa complaisance passive, en utilisant a l'etranger la puissance de la maison, qu'il allait jusqu'a compromettre, dans ses arbitrages; et Sedille lui-meme, le marchand de soie, ebranle a la suite d'une liquidation terrible, s'etait fait preter une grosse somme, qu'il n'avait pu rendre. Seul, Daigremont gardait son independence absolue vis-a-vis de Saccard; ce qui inquietait ce dernier, parfois, bien que l'aimable homme restat charmant, l'invitant a ses fetes, signant tout lui aussi sans observation, avec sa bonne grace de Parisien sceptique qui trouve que tout va bien, tant qu'il gagne." (Zola, L'Argent, p. 310-1)

"Saccard had succeeded in getting hold of all the members of the board of directors, in buying them out literally, in most of the cases. It is due to him, that the marquis de Bohain, compromised in a story of bribing equivalent to a swindle, discovered with his hand in the bag, could escape from a scandal, by compensating the robbed company; and he became subsequently his humble servant, while remaining with his head high, an aristocrat, the best ornament of the board. Huret, as well, since Rougon has dismissed him, after the theft of the wire that announced the transfer of Venetia, has committed himself fully to the success of the Universelle, representing it at the Parliament, fishing for it in the dirty waters of politics, keeping the largest part of the shameless scams, that could throw him one day to prison.

And the vicomte de Robin-Chagot, the vice-president, received a hundred thousand francs as a secret fee for signing without examination during the long absences of Hamelin; and the banker Kolb was paid also for his passive readiness to oblige, while utilizing abroad the strength of the company, which put it even in jeopardy in his arbitrations; and Sedille himself, the silk merchant, undermined by the consequences of a terrible liquidation, was lent a huge sum, that he was unable to reimburse. Only, Daigremont kept his full independence toward Saccard; which bothered the latter, sometimes, although the nice person remained charming, inviting him to his feasts, signing everything without inquiring, with his amiability of a skeptical Parisian that finds that all is well, as long as he is gaining money."

Minority shareholders themselves have today a distribution that varies significantly from the past. The institutional shareholders have, according to Monks, 47.4 percent of the capital of the American corporations, $4.35 trillion in 1996, 57 percent of the capital of the 1,000 largest companies, and half of this capital or 30 percent of the whole capital is held by public funds or pension funds. "In mutual funds (more formally known as investment companies), the 'independent directors' are chosen under the provisions of the federal Investment Company Act of 1940. They are paid extremely well for services that basically consist of deciding whether to ratify the investment management contract (with a firm whose principals invited them to serve as directors), and they almost invariably vote to do so. In other words, mutual fund trustees are paid so much too much for doing so little that they are unlikely to disturb their sponsors." (Monks, The Emperor's Nightingale, p.148) The fiduciaries of the funds must not be nominated and paid by the companies that they are supposed to control. We shall see in the cases analysis how those fiduciaries behave in cases of abusing the rights of the minority shareholders and what is the level of their courage and integrity.

A basic factor in the need of the preponderance of ethics over the law is the ignorance of many shareholders of basic terms in the prospectus of companies, which are for them like Chinese. The law and the SEC regulations maintain that if all the important issues are disclosed in the prospectus - the companies have performed legally, even if the most important issues are disclosed in such a way that it is almost impossible to notice or understand them, as we shall see in the empirical part of the book.

Furthermore, even according to GAAP's rules, a company can attribute 'extraordinary' costs, due to a restructuring or purchase of a company, whose main assets are intangible, as costs which are treated separately in the financial statements, and which analysts do not take usually into consideration in the valuation of the company. This gives the possibility to companies and to those who control them to do whatever they like in the financial statements and in the prospectuses, while strictly obeying the regulations of the SEC and of GAAP.

Minority shareholders, and especially small investors, who do not understand anything in these intricacies, buy the shares at inflated prices at the stock exchange or at a shares' offering, and often the shares subsequently collapse, while the company has not committed any illegal act. The SEC has decided to change its rules and asks now from the companies to publish a prospectus in a comprehensible language to the average stockholder, and in parallel the rules of the financial reports on the extraordinary costs are being revised. Those changes are done due to the fact that according to Compustat for the US industrial companies, the value of the tangible assets amounted to 62 percent of the market value in 1982, while in 1992 it amounted only to 38 percent! The repercussions of this state of affairs, which are extremely dangerous for minority shareholders, is examined at length in the case study of the American company in the empirical part of the book.

We have to define the legal term of minority shareholders, as it is used in this book. A minority shareholder is defined as a shareholder who does not exert control over a company. The majority shareholders almost always exert an absolute control over the company, its management, its board of directors, and so on. But there are many companies that are controlled by shareholders who own only 40 percent, 30 percent, 20 percent, or less of the shares, and whom however exert full control over the company, as the remainder of the shares are scattered among a large number of shareholders, with every one of them having a minimal percentage being unable to gather a number of shares which is similar to those of the majority shareholders. In this event, all minority shareholders who are scattered, although together they could control even 80 percent of the shares, are defined as minority shareholders, as every one of them is a minority shareholder, and they cannot assemble enough votes to act as majority shareholders.

There are also cases where there are two or three groups of shareholders, with every one of them having 10 percent or 20 percent of the shares, and who are minority shareholders. They can elect their members to the board of directors and split the control or they can make coalitions between two groups of shareholders against two others, etc. Here also, those who control the company are the 'majority' shareholders, as they have the majority of the seats in the boards of directors, even if in reality they have less than 50 percent of the shares of the company, while those who do not control the board of directors are defined as minority shareholders even if they own together the majority of the shares. In many cases, the shares are distributed among a large number of shareholders who own each a few percentages, one percent, or even less of the shares. In those cases, or if the managers own themselves a few percentages of the shares, the management of the company manages often to get the control of the company and of the board of directors, and they can do what they wish in the company, as the shareholders are too scattered and cannot exert their power.

The definition of minority shareholders in this book will be therefore shareholders who do not exert control over the board of directors of the companies, even if together they own the majority of shares, and the majority shareholders are defined as those who control the board of directors of companies, even if effectively they own much less than the majority of the shares. The analogy between this situation and the political system of nations is clear. Companies are still at the stage of oligarchies and have not reached the status of democracies.

As far as the author of this book could analyze, most of the public companies traded in the stock exchanges of the US, France and Israel, are controlled by groups of shareholders who own less than 50 percent of the shares of the companies. If the minority shareholders who are effectively the majority would be conscious of their power, and if the boards would be elected only in proportion to the ownership while the remainder of the members would be elected by activist associations, this could revolutionize the modern business world, safeguard the rights of minority shareholders, and prevent the abuse of the shareholders by oligarchies backed by the executives of the companies.

The 'proletariat' of the shareholders, who are not organized, are too often abused, and the time is appropriate for them to get organized directly or through the activist associations, in order to exert their legitimate power and preserve their rights. There is no reason whatsoever that the last vestige of oligarchies, the business world, would remain immune to the democratic evolutions and revolutions that prevail nowadays throughout most of the countries of the world.

The evolution toward participation in the control of companies by minority shareholders is in progress, although very slow, but nevertheless we could notice a tendency, which is reinforced every day. "The California Public Employees Retirement System, the New York State Common Retirement Fund, and the Connecticut State Treasurer's Office have jointly pressured several dozen firms to put a majority of outside directors on their boards' nominating committees... In the future, major shareholders will include employees as well as institutional investors... we may even witness a general restructuring in corporate ownership, one that induces managers to shift their allegiance from the wealthy to the less advantaged: Pension funds and other institutional investors already account for approximately 40 percent of the shares traded, with 10 percent of the nation's households commanding most of the rest... the demand for a global managerial ethics will become increasingly urgent. American managers will have to compete not only on the basis of technique but of democratic values as well." (Kaufman, Managers vs. Owners, p.196-8)

There is a difference in the modes of operation of the stock exchanges in the world. In Great Britain, for example, the participation in the capital is much more concentrated and institutionalized than in the United States. The shares' issues are principally offered to the existing shareholders in order to permit them not to dilute their ownership. Nevertheless, the basic ethical principles of the financial markets are identical. The just transactions should be performed out of free will, it is impossible to force a shareholder to buy or exchange a share against his will. The transactions should be done for the good of both parties, it is impossible to base a transaction on the oppression of part of the shareholders, and they should be based on information, which is common to all the shareholders. Insider trading is therefore strictly prohibited as it favors only a part of the shareholders to the detriment of those who do not possess the information.

The class actions are very limited in their scope, rewards and efficiency. They are time consuming, and some people even alleged that they benefit mostly the lawyers that handle the cases. Still, until more efficient vehicles are devised, many shareholders resort to class actions. The empirical part of the book has many references of class actions. A detailed explanation of the process of class actions is given at the end of the book.

The origin of the abuse of minority shareholders comes mainly from the greed of some of the majority shareholders, who in some cases has no limit. Those majority shareholders believe that they can do anything, risk more and more, since they find themselves unpunished, while remaining within the very large margins of the law. The minority shareholders who are wronged do not learn the lesson and continue to invest in companies which are conducted in an unethical manner. This is why it is needed to examine in depth the legal protection of those minority shareholders and its efficiency, in order to verify if the law suffices for their protection, or if the minority shareholders need an ethical protection, which has a much wider scope.

3
THE ATTITUDE OF SOCIETY

"All truth passes through three stages. First, it is ridiculed. Second, it is violently opposed. Third, it's accepted as being self-evident."
(Arthur Schopenhauer, German philosopher, 1788-1860)

Members of society have a tendency to overlook events that do not concern them directly, and it is against this indifference that one has to fight, as an immoral ambiance has a tendency to penetrate to all domains thus affecting all members of society. One is always a client, or a minority shareholder, or a supplier, or at least a member of society, who is affected by ecological crimes or others. An immoral ambiance will make all of us victims, exactly like a totalitarian regime turns ultimately against the majority of its citizens.

Who then is fit to speak with authority about business ethics and how the environment of the businessmen in general and the minority shareholders in particular will receive their ideas and recommendations? Would it be businessmen who are also part of the academic world or who are simply humanists, erudite in philosophical, historical and literary texts? The moralists of business are more and more convinced that a combination of all those qualities would be optimal to deal with business ethics.

"In his introduction to Business as a Humanity, Thomas J. Donaldson says that the authors of this volume agree that humanities' texts, e.g., philosophical, historical, and literary works, should be assigned in business schools. Business ethicists have shown the importance of philosophical and historical research in business ethics. Possibly, however, not enough has been said about the importance of literature in business ethics. In his Business as a Humanity: A Contradiction in Terms? Richard T. De George maintains that Literature offers the business student 'subtlety of insight, beauty of language, imagination, and vivid description that puts most texts to shame… (Students) do need to understand people and their motives, to know how to read and judge character, and to have the ability to imagine themselves in another's shoes, be they those of a competitor, a boss, or a subordinate. For those dedicated to the case method, novels, short stories, and plays offer an inexhaustible storehouse of riches, more detailed, subtle, and complete than most cases written up for courses." (Business Ethics Quarterly, January 1998, Klein, Don Quixote and the Problem of Idealism and Realism in Business Ethics, p. 43)

This book fully adheres to the principles mentioned above by two of the most prominent business ethicists, Donaldson and De George. It juxtaposes plays by Miller, Pagnol, Brecht and others, novels by Zola, Pagnol, Cervantes, Kafka, and others, fables, poetry, etc., with ethical situations, trying to find analogies between the imaginary and real situations emphasizing the ethical dilemmas of the novel heroes and the modern businessmen. Furthermore, the book is based on a thorough study of cases which enables detailed judgement of the situations, persons and psychological conduct of the ethical heroes or villains.

As psychology is at the basis of ethical conduct in business, we cannot understand the conduct of the businessmen without analyzing in depth their character and motives. But is it practical to base the ethical principles on philosophical, religious or literary bases? Do we not incur the risk to be treated as Don Quixote, who was completely subjugated by his ideals? Can we be practical, succeed in business and retain however the ideological and literary bases? Would the environment of the businessmen treat us with respect, commiseration, alienation or envy? This is the basic dilemma of many businessmen who try to reconcile the ideal and the reality without becoming a Don Quixote.

"Cervantes condemns the books of chivalry, as embodied in his character Don Quixote, as both fantastical and dangerous. The chivalric hero may seduce people into believing that the improbable can be achieved with ease. Cervantes' character, Don Quixote, shows that this is not the case. Here is a hero possessed of fine qualities of both character and intellect who sallies forth in the name of justice and human betterment. Nonetheless, while being inspired by high ideals, his efforts are futile because he pays little or no attention to the means necessary for achieving these ends, and he fails to gain requisite knowledge of the circumstance and conditions necessary to properly understand human actions. Cervantes seems to be saying that when idealistic theory is divorced from practice, however noble the theory and good the intentions, requisite skill, judgment, and discretion will be lacking and the human good will not be advanced. (Business Ethics Quarterly, January 1998, Klein, Don Quixote and the Problem of Idealism and Realism in Business Ethics, p. 44)

"So far our Don Quixote scenario could provide a cautionary tale for business ethics. Some businesspeople with a good deal of practical experience have looked askance at the sallies of philosophical bookish knights armed with their (e.g. deontological and/or utilitarian) moral theories which they learned 'living in the books'. They might argue that there is something comic in some philosophers' attempts to solve the morally complex problems of business by applying moral theories to overly simplified 'case studies'. (Business Ethics

Quarterly, January 1998, Klein, Don Quixote and the Problem of Idealism and Realism in Business Ethics, p. 45)

The environment of the ethical businessmen or people in general can treat them as courageous, crazy or impertinent, as is maintained by Sancho Panza or as virtuous but calumniated as maintained by Don Quixote:

"En lo que toca – prosiguio Sancho – a la valentia, cortesia, hazanias y asunto de vuestra merced, hay diferentes opiniones: unos dicen: 'Loco, pero gracioso'; otros, 'Valiente, pero desgraciado'; otros, 'Cortes, pero impertinente'; y por aqui van discurriendo en tantas cosas, que ni a vuestras merced ni a mi nos dejan hueso sano.

Mira, Sancho – dijo don Quijote – donde quiera que esta la virtud en eminente grado, es perseguida. Pocos o ninguno de los famosos varones que pasaron dejo de ser calumniado de la malicia." (Cervantes, Don Quijote de la Mancha II, p. 43)

"In what pertains, continued Sancho, to courage, courtesy, exploits, and business of your grace, there are diverging opinions: the ones say: 'Crazy, but gracious'; the others, 'Courageous, but unhappy', others, 'Courteous, but impertinent' and from there they discuss so many things, that neither to your grace neither to me they leave a whole bone.

- Look there, Sancho – said don Quijote – in the place where virtue exists at a large degree, it is persecuted. A few or none of the respectable and famous men who have existed have escaped from the calumny of malice."

And Peters and Waterman reinforce the importance of the moral element in our life by affirming: "We desperately need meaning in our lives and will sacrifice a great deal to institutions that will provide meaning for us." (Peters and Waterman, In Search of Excellence, p. 56) And they continue: "an effective leader must be the master of two ends of the spectrum: ideas at the highest level of abstraction and actions at the most mundane level of details." (same, p. 287) And thus, like Don Quixote, the leader has to possess a vision: "Attention to ideas – pathfinding and soaring visions – would seem to suggest rare, imposing men writing on stone tablets." (same, p.287)

Ibsen illustrates in a dramatic way the ethical dilemma of Dr. Stockman, the officer of the municipal Baths, who has discovered that the water of the Baths is polluted, and announces it publicly at the risk of alienating himself from his whole town, which could be ruined as a result of his discovery. He is indeed called The Enemy of the Public, dismissed from his job and ostracized by his community. In a decisive confrontation with the citizens' assembly, Dr. Stockman maintains that the majority has not the monopoly over truth and morality, and he advocates with vehemence the right of the minority to embrace the truth, which can be opposed to that of the majority, but which nevertheless is the unique moral truth, over which he will fight without heeding the consequences. Stockman, the individualist, who fights alone against everybody else, has even a predestined name very relevant to this book, as he is called stock-man, the man with a stock, the individual shareholder.

"I propose to raise a revolution against the lie that the majority has the monopoly of the truth. What sort of truths are they that the majority usually supports? They are truths that are of such advanced age that they are beginning to brake up. And if a truth is as old as that, it is also in a fair way to become a lie, gentlemen. (Laughter and mocking cries.) Yes, believe me or not as you like; but truths are by no means as long-lived as Methuselah – as some folk imagine. A normally constituted truth lives, let us say, as a rule seventeen or eighteen, or at most twenty years; seldom longer. But truths as aged as that are always worn frightfully thin, and nevertheless it is only then that the majority recognizes them and recommends them to the community as wholesome moral nourishment. These 'majority truths' are like last year's cured meat – like rancid, tainted ham; and they are the origin of the moral scurvy that is rampant in our communities." (Ibsen, An Enemy of the People, p. 256-7)

If the majority of businessmen maintains that you cannot argue with success and that everything is permitted to obtain this success, there could still exist a minority that maintains that the absolute value is ethics and it is despicable to succeed by despoiling the rights of minority shareholders, stakeholders and, ultimately, everybody. The author of this book believes that this minority is probably right. They will ridicule us as they have done to Don Quixote, they will fight us as they have done to The Enemy of the People, but finally, the truth of the minority will be perceived as self-evident, as democracy, as Human Rights, as equality of mankind, black, yellow or white, men and women, Christians, Moslems or Jews, Americans, French, British, Dutch or Israelis.

At the third act of Marcel Pagnol's Topaze we discover that the honest teacher was transfigured and has become corrupted. He is sitting behind a desk, while on the walls we can read: 'Soyez brefs' – be brief, 'Le temps, c'est de l'argent' – time is money, 'Parlez de chiffres' – speak in numbers. Topaze is a frontman, a man of straw. He feels soiled and cannot suffer the look of an honest man. He tries to maintain still that money does not bring happiness, but Suzy, the woman he loves answers him 'No, but it buys it from those who make it'. In the corrupted environment he starts to prove himself and becomes much more competent than his colleagues. In confrontation with his old friend he justifies himself: 'Tout ce que j'ai fait jusqu'ici tombe sous le coup de la loi. Si la societe etait bien faite, je serais en prison.' – 'All that I have done is legal. If society was just, I would have been in prison.'

And he concludes: "Regarde ces billets de banque, ils peuvent tenir dans ma poche mais ils prendront la forme et la couleur de mon desir. Confort, beaute, sante, amour, honneurs, puissance, je tiens tout cela dans ma main... Tu t'effares, mon pauvre Tamise, mais je vais te dire un secret: malgre les reveurs, malgre les poetes et peut-etre malgre mon coeur, j'ai appris la grande lecon: Tamise, les hommes ne sont pas bons. C'est la force qui gouverne le monde, et ces petits rectangles de papier bruissant, voila la forme moderne de la force. (Pagnol, Oeuvres Completes I, Topaze, p. 453)   "Look at those banknotes, they can fit in my pocket but they will soon take form and color of my desire. Comfort, beauty, health, love, honors, power, I hold all this in my hand... You are bewildered, my poor Tamise, but I will tell you a secret: in spite of the dreamers, in spite of the poets and maybe in spite of my heart, I have learned the big lesson: Tamise, men are not good. It is power which governs the world, and this small rectangles of noisy paper, this is the modern structure of power."

Pagnol, alternatively pessimist and optimist, describes to us admirably the dilemmas of all of us and how many of us resolve them. If Topaze would have remained in his environment, as a teacher with an honest headmaster, he would have remained the most honest man. But it is because he has suffered injustice and has joined a corrupted society that he has been corrupted himself and has sold his soul, while being convinced that he is on the right track. He becomes much more corrupt than his mentors, as he thinks that this is the only way to survive, and he finds justifications that manage to convince him as well. This is therefore the predominant role of the moral environment, which succeeds in most of the cases, especially with men who do not have a strong and well-formed character, to fashion its member into its image. Tell me who your friends are, and I will tell you who you are.

4

THE EXCESSIVE PRIVILEGES OF THE MAJORITY SHAREHOLDERS

"I've become rich, friendless and mean,
and in America, that's as far as you can go."
(Mr. Vandergelder, "Hello, Dolly!")

We live in a time of mergers and acquisitions, and in many cases, the shareholders having the control of the companies with 30 percent or 40 percent of the shares want to merge with another company or privatize their company by forcing the other shareholders to agree to a takeover bid. This bid is done in most cases when the shares' price is very low, after having collapsed due to market conditions, unexpected bad financial results or indirect manipulation of the shares' prices.

One should not forget that the market heavily penalizes companies that fall short of obtaining their forecasted results. We can imagine that the CEO of a company with the collaboration of the majority shareholders, who decide on his remuneration, give a growth forecast of 50 percent, which is much higher than the actual growth. The analysts take this growth in consideration and give the valuation of the company a high multiple of the current profitability that can reach even 100. The company makes a public offering and the majority shareholders who are insiders sell part of their shares, for example 10 percent. If later on, the growth is much less than forecasted and the company expects losses, the shares' price may collapse by even 90 percent.

The controlling shareholders or their associates make a takeover bid for all the shares at the minimum price of 10 percent in our example, and they buy all the shares with the amount raised at the maximum price in the shares' offering. From the moment that the company is privatized it can again reach profitability or have a very high growth rate, except that by then the minority shareholders will not benefit from the turnaround and the increase in valuation of the company, as they were forced to sell their shares at the minimum price. Those examples are very common as we are going to learn from the analysis of the cases in this book.

"Insider trading, or the use of insider information, represents a special case within this category: it involves the use of confidential information about the firm's future performances by the employee on the financial market, in order to realize a speculative gain for himself or for some third party. Such practices are to be condemned for two reasons:

1 - The employer is in fact robbing his or her own firm, since she could only receive the crucial information as a member of it. Moreover, he or she only received this information under the condition that s/he would use it to serve the corporate interest.

2 - Third parties have also been damaged: those agents who were deprived of the information while dealing on the same financial markets or contracting with the same firm will unjustly suffer losses as a consequence of the practice."

(Harvey, Business Ethics, A European Approach, Gerwen van, Employers' and employees' rights and duties, p.81)

Insider trading is surely not a modern invention. Zola described it brilliantly in L'Argent – The Money, where Saccard and his colleagues commit insider trading and speculations to the detriment of the minority shareholders and remain practically unpunished. "L'Argent serait-il donc un conte moral ou les mechants sont punis et les bons recompenses? Bien sur, l'escroc Saccard est emprisonne – pas pour longtemps. Mais le 'filou' Sabatini, l' 'adroit' Nathanson et le malhonnete Fayeux courent encore. Et surtout beaucoup de gens honnetes dont la seule erreur a ete leur pitoyable naivete restent des victimes. C'est le cas de l'agent de change Mazaud mais surtout de tous les petits actionnaires. Les gros s'en tirent mieux. Si la justice n'est pas retablie par la condamnation effective des profiteurs dans la diegese elle-meme, du moins l'est-elle par leur condamnation verbale." (Commentaires par Therese Ioos, Zola, l'Argent, p. 502)

"Is L'Argent a moral tale where the bad people are punished and the good ones rewarded? Of course, the swindler Saccard is imprisoned – not for long. But the 'crook' Sabatini, the 'skillful' Nathanson and the dishonest Fayeux are still at large. And especially many honest people whose only mistake was their pitiful naivete remain their victims. It is the case of the broker Mazaud but especially of all the small minority shareholders. The big ones succeed more. If justice is not reestablished by the effective condemnation of the profiteers in the story, at least it is done in their verbal condemnation."

Guido Corbetta, in one of the rare articles on the ethical questions in the relations between companies and shareholders divides the most common forms of ownership of medium-sized and large companies in four categories:

"1. Family-based capitalism: ownership is concentrated in the hands of one or a few families, which are frequently related to one another. Sometimes one or more members of the family is directly involved in running the company... This form of ownership is particularly common in Italy, but there are large family businesses practically everywhere.

2. Financial capitalism: ownership is concentrated in the hands of one or just a few private and public financial institutions which, through a system of cross-holdings, control companies and intervene in their management... Ownership also implies powers to appoint management and steer corporate strategy... This form of ownership (with some slight differences) prevails in Germany, Japan and some other countries like Holland and Switzerland; it is rapidly becoming more common in France too.

3. Managerial capitalism: ownership is shared among numerous stockholders, none of whom exercises any significant control over the activity of the managers who run the companies. The management of these companies therefore becomes a kind of self-regenerating structure... It is particularly important in the Anglo-American business world.

4. State capitalism: through central and peripheral agencies or corporations set up ad hoc (as in the case of, for instance, IRI and ENI in Italy), the state has direct control over the companies. The existence of this form of capitalism clearly stems from a certain view of state intervention in the economy. In Italy, France and Spain there are major groups belonging to this category...

In cases of family-based capitalism and financial capitalism, for example, boards of directors are appointed by the majority shareholder or by a coalition of shareholders who are often themselves members of the boards, which appear to be the real organs of corporate governance. In cases of managerial capitalism, board members are instead 'co-opted' by the management itself. Save a few noteworthy exceptions, the choice falls on people whose most important characteristic appears to be their willingness to endorse without question whatever proposals the top managers who are also board members may submit. The board of directors thus eventually loses its role as collective organ of corporate governance and often becomes a false front used to give greater authority to decisions made by others."
(Harvey, Business Ethics, A European Approach, Corbetta, Shareholders, p.89-90)

We have dealt at length throughout this book on the differences between the different types of shareholders, especially the majority or controlling shareholders who are called in Corbetta's article the 'governor' shareholders and the minority or small shareholders who are called in Corbetta's article the 'investor' shareholders. The characteristics of both categories are summarized as follows:
"We define our shareholder as a 'governor' when:
- the percentage share of capital stock owned is high;
- development of the firm is substantially dependent on the economic resources made available by the shareholder and, likewise, the economic fortunes of the shareholder depend significantly on the firm's profitability;

- the shareholder exercises his or her power to intervene in decision-making processes by appointing the firm's management, steering corporate strategy and monitoring and appraising the management's performance;
- any decision to sell the shareholding is limited by sentimental reasons, in the case of family businesses, or by complex strategic assessments which may occasionally even have implications for national equilibrium (as was recently the case with operations conducted in Germany and France).

We define the shareholder as an 'investor' when:
- the (percentage) share of capital stock owned is small, often a fraction of a percentage point;
- the link between the development and profitability of the firm and the fortunes of the shareholder is not very close: the company gathers its resources from a very large number of shareholders, each of whom makes only a limited contribution to the firm's needs; likewise, the income of each individual shareholder does not come from the dividends distributed by the firm;
- there is little likelihood that shareholders' opinions about management appointments and corporate strategy will influence decisions. On a practical level a 'shareholders' democracy' – i.e. effective control over management by numerous small shareholders – is not feasible;
- the decision to sell the shareholding is taken only on the basis of assessments of returns. 'Abandoning' is often preferable to 'expressing dissent' and, even more so, to 'remaining bound'."
(Harvey, Business Ethics, A European Approach, Corbetta, Shareholders, p.92)

The management of management-controlled companies are reluctant to hand over many of their autonomy to the shareholders. This increases the possibility of anti-company behavior on the part of the managers, who are concerned only with getting the maximum personal gain even when this puts the very survival of the company in jeopardy. Corbetta concludes that the governor-shareholder is not morally justified in using the company for his own ends, not even considering that his own compensation is secondary to that of other stakeholders. This article summarizes in a very efficient way all the analysis of the struggle for power and the different sets of interests between the majority and minority shareholders, and emphasizes the risks that the small shareholders incur from not controlling in fact the companies, thus enabling the majority shareholders to misuse their power and to wrong the other shareholders.

It is time to describe in a few words the evolution of the control of companies in Israel. The state of Israel has come full circle in its first 50 years of establishment. In its early years (1948-1968), the economy of Israel was based on companies owned by families and by the Histadrut, the workers' syndicate, and was controlled by a very small number of people. The economy has evolved to companies owned by financial institutions and by the government (1968-1990), but in the last decade of the century, due to privatizations, mergers, bankruptcy of many organizations, and the decline of the Histadrut, the largest part of the economy is effectively controlled by 10 to 20 wealthy Israeli and Diaspora-Jewish families. The Israeli economy is today an oligarchy, which could very well determine the economic fate of the country and possibly, in the future, its political fate in a small boardroom of 40 square meters.

The author of this book is convinced that the present state of affairs regarding minority shareholders in France, Israel and the United States is to their detriment, in all the possible contexts. In family-owned companies, they cannot influence the decisions which are taken by the 'Grandes Familles', the richest families, and which favor uniquely those families and rarely the other shareholders. The members of the family are elected to the key managerial positions in the companies, even if they are incompetent, the families do all that is necessary to keep their effective control over the companies, even if it is to the detriment of those companies. As the families have many ramifications to their investments, they can cause the collapse of the price of the shares in one company and enable another company to buy it for an extremely low price.

The 'governors' are convinced that if they are strongly involved with the companies, they control it and they supply it with funds, they have the right to do whatever they want with 'their' companies, and the minority shareholders are treated like speculators, who are not interested in the well-being of the companies but rather in a quick return on their investment. Even if this is true in certain cases, this does not decrease the rights of the minority shareholders, who are in many cases interested in the fate of the company no less than its governors. The cases of the managerial companies are even more dangerous for minority shareholders as the directors jeopardize the company itself in order to increase their personal benefits.

The democracy of the shareholders is completely utopic, the shareholders can shout, protest, be indignant, criticize or threaten on the Internet or in the shareholders' meetings, yet their influence is in most cases nil in all categories of the companies. This is the reason why they have to obtain new rights, even if they do not request it yet. In many cases the minority shareholders collaborate unknowingly with the majority shareholders in order to despoil their own rights. They have the opportunity to participate in shareholders' meetings, which are in many instances a ridiculous circus, manipulated very skillfully by the majority shareholders, who are assisted by the management of the companies.

And even if they participate in the meetings, which is very rare, they have no chance to win against the oiled machine of the owners who control the companies. The cases described in this book illustrate those statements and show how it is possible to eliminate from the protocol touchy questions and answers to minority shareholders, how is it possible to treat as a ridiculous Cassandra troublemakers who disclose the schemes of the owners, thus even augmenting the adhesion of the other shareholders, and how ultimately the minority shareholders cooperate unknowingly or against their wishes in the schemes of the majority shareholders.

The majority shareholders manipulate the greed of the other shareholders who want to win the jackpot at all cost in the stock exchange and make them lose in many cases all their investments. One would think that he is at the court of Mantova, where the masked Rigoletto assists the abductors of his daughter Gilda, without hearing or seeing that they abducted his daughter. They take Gilda to the duke who rapes her, and instead of avenging herself, Gilda lets herself be murdered by Sparafucile in order to save the duke. It is Rigoletto who is punished instead of the duke. Monterone who is imprisoned for having insulted the duke who abducted his daughter too, summarizes the dilemma of the weak toward the mighty: "Poiche fosti invano da me maledetto, ne un fulmino o un ferro colpiva il tuo petto, felice pur anco, o Duca, vivrai." (Piave, Rigoletto, Act III, p.14) "And since my curse has left you unharmed, and no lightning or iron has cracked you skull, you will even though live happily."

The collaboration of the victim with the aggressor is a well-known psychological fact, but the purpose of this book is to eradicate this mentality which is too widespread, by eliminating the excessive rights of princes, dukes or majority shareholders to the detriment of the minority shareholders. The modern democratic evolution should not stop at the door of the business world. The kings do not amuse themselves anymore, as in Le Roi S'amuse of Victor Hugo, adapted to the opera Rigoletto by Piave, the tyrants have disappeared in most countries, it is high time that the 'droits du seigneur' of 'first night privileges' will disappear from the Medieval courts of the companies as they have disappeared from the court of the duke of Mantova.

Milken, the indisputable hero of the financial world of the '80s, perceived himself as above the legal and moral constraints and thought that they were good only for the 'footsoldiers' – in our case the minority shareholders, the less influential, the less creative, less aggressive, less visionary. There are therefore double standards for the footsoldiers and for the Knights, just as in the Middle Ages. This is the core of this book, how to evolve from the dark and unhealthy epoch of the Middle Ages, where a large part of the business world is still wallowing, to the Renaissance period of the years 2000, and to have the same standards for minority shareholders, as were achieved for minorities all over the civilized world, by Human Rights, the welfare society and democracy. Time is of the essence, as the situation is getting worse instead of improving.

The world economy becomes more and more concentrated in the hands of a small number of huge organizations, which control the economy, without being adequately controlled by the governments and the citizens, and least of all by the shareholders. In 1994, 1,300 companies have participated in mergers amounting to $339 billions. And today the mergers are even larger. The modern empires of companies are much more influential than the monopolies of the Carnegies and the Mellons. The profits of Wall Street in the last years of the century were stunning. The volume of the financial transactions of the `90s is 40 times higher than the productive economy of the US, while the volume of transactions of CS First Boston is higher than the GNP of the US. The SEC has not the necessary funds to control effectively those giants and the only safeguard against them is ethics.

Are all business ethicists preaching in the desert or is the majority of the population really conscious of the serious situation which predominates in the business world? What is very indicative in this respect is the level of trust of Europeans toward the institutions, like Church, the Army, Education, Law, Press, Trade Unions, Police, Parliament, Civil Service, Social Security, the EC and NATO. The most striking feature is that the level of mistrust or lack of confidence of the French people toward the Major Companies is only 30 percent, the lowest in all Europe. 70 percent of the French believe in Major Companies! The level of lack of confidence towards those companies is 37 percent in Italy, 47 percent in Ireland, 49 percent in Belgium, 50 percent in Spain and Great Britain, 51 percent in the Netherlands, 53 percent in Portugal, 62 percent in Germany, and on the average is – 47 percent. Surprisingly enough the British are much less credulous than the French toward the Major Companies and the most mistrusting are the Germans with a level of mistrust of 62 percent. This is completely opposite to the preconceived ideas about the Europeans, or it may indicate that the French Major Companies are much more trustworthy than the British or German ones...

Majority shareholders, executives and members of the boards of directors benefit from insider information, which is not accessible to minority shareholders. If the insiders utilize this information to buy or refrain from buying shares of the companies, they commit a despoliation of the rights of the minority shareholders. They risk nothing in buying the shares, as they know in advance that their prices will increase as a result of good financial results, a merger or a scientific discovery. On the contrary, if they sell their shares before the publication of negative financial results, they do not incur losses from the collapse of the shares' price. The empirical part of the book will demonstrate how the insiders benefit from such information and remain unpunished, as it is almost impossible to prove an abuse of insider information.

"The game, then, like the manipulated market that is the outcome, is unfair – unfair to some of the players and those they represent – unfair not only because some of the players are not privy to the most important rules, but also because these 'special' rules are illegal so that they are adopted only by a few of even the privileged players." (Rae, Beyond Integrity, Werhane, The Ethics of Insider Trading, p. 518) Even worse, the insiders register their companies in Delaware, which enables them to benefit from a complete freedom of action in the governance of their companies. "Delaware, for example, has few constraints in its rules on corporate charters and hence provides much contractual freedom for shareholders. William L. Cary, former chairman of the Securities and Exchange Commission, has criticized Delaware and argued that the state is leading a 'movement towards the least common denominator' and 'winning a race for the bottom'." (Rae, Beyond Integrity, Jensen, Takeovers: Folklore and Science, p. 530)

If this is the case, does the SEC advise the shareholders of the risks that they incur when they buy shares of companies registered in Delaware? Does it try to change the corporate laws of this state? In a case of this book we shall see how majority shareholders have rendered almost impossible an organization of minority shareholders against their wrongdoing by relying on the corporate laws of the State of Delaware and how the SEC has not done anything to remedy the situation.

The present state of affairs is unfortunately like in the Fables of Aesop and La Fontaine, as human nature has not changed since those ancient times. The mighty always find reasons to abuse the rights of the weak - weird, legitimate or even moral. This is why there is a constant abuse of the rights of the weak by the powerful, and the weak have to suffer the consequences of their 'crimes', as they trouble the water of the wolves, they speak ill of them, and they have too many brothers. In order to punish their crime to want to drink in the same waters as the wolves, they almost always lose, as they are allowed to invest their money but they are prohibited from sharing the profits with the mighty. This is why they almost always lose in court, and they are even fined for the "arrogance" they showed in trying to sue the powerful, as in the case of the French company in this book.

La Fontaine illustrates this state of affairs in the moral of his fable Le Loup et L'Agneau – The Wolf and the Lamb: 'La raison du plus fort est toujours la meilleure' – Might is Right. The fable of Aesop, which La Fontaine adapted, can summarize in the best way this chapter: "Wolf, meeting with a Lamb astray from the fold, resolved not to lay violent hands on him, but to find some plea to justify to the Lamb the Wolf's right to eat him. He thus addressed him: 'Sirrah, last year you grossly insulted me.' 'Indeed', bleated the Lamb in a mournful tone of voice, 'I was not then born.' Then said the Wolf, 'You feed in my pasture.' 'No, good sir,' replied the Lamb, 'I have not yet tasted grass.' Again said the Wolf, 'You drink of my well.' 'No,' exclaimed the Lamb, 'I never yet drank water, for as yet my mother's milk is both food and drink to me.' Upon which the wolf seized him and ate him up, saying, 'Well! I won't remain supperless, even though you refute every one of my imputations.' The tyrant will always find a pretext for his tyranny." (Aesop's Fables, The Wolf and the Lamb)

5

INTERNET AND TRANSPARENCY AS ETHICAL VEHICLES

"The accomplice of a thief is his own enemy;
He is put under oath and dare not testify."
(The Bible, Proverbs, 29:24)

The activists shareholders, who are more and more influential, can communicate via the Internet, which enables free, instantaneous, interactive communication between shareholders, between shareholders and companies, and between shareholders and the organizations that are supposed to safeguard their interests as the members of the board of directors, independent directors, fiduciaries, the SEC, etc. In the future, they would be able to ratify decisions that will be submitted to them via the Internet, receive all the required information and financial reports for their decisions from the Internet, and obtain answers to their queries very promptly.

In the business world, as in the political and social world, the tendency is for everybody to mind their own business, and even if the rights of others are wronged they seldom interfere, as they do not want to make enemies, they do not have time for such occupations, or "they didn't help me when I was in need so why should I help them now?" etc. But if it is possible to denounce the crimes without being discovered, there is a tendency to do so, in order to have a clean conscience. The Internet is the best vehicle to do so as it enables you to retain your anonymity while disclosing to the whole world the facts that prior to then were hidden. Light is the worst enemy of criminals who prefer to work in the dark. In some business circles the law of Omerta (Silence, like in the Mafia) prevails, and rarely does someone dare to transgress this law. But the Internet changes this setup, as the whistle-blowers remain concealed and the truth is revealed.

Unfortunately, it is possible to utilize this vehicle also to defame businessmen and companies, manipulate shares, spread rumors and misinform the shareholders by interested parties – the companies, the majority or minority shareholders, competition, or others. As everyone keeps his anonymity, they remain unpunished, although there are some attempts to raise the curtain over those people in extreme cases. Misinformation or not, the minority shareholder has at least the opportunity to be informed about unethical acts performed by the companies or to denounce them in advance. He has only to discern the true and false information, which is better than before when he had no access to the true information.

The ideal would be that companies would be transparent to the shareholders and that all the shareholders would receive simultaneously the same information, whether they are minority or majority shareholders. No more insider information, no more abuse at the detriment of shareholders who live far from the headquarters of the company and who have no access to the information divulged by the insiders to the boards of directors. We could also imagine a black list, established by activist associations and published on the Internet, of companies and persons who do not behave ethically, who went bankrupt, who were condemned by the courts. Accessible to everyone around the world, this list could induce the companies and their executives to conduct themselves ethically and legally, make their utmost effort not to go bankrupt and to repay their debts even if they do not have a legal obligation to do so. It would be recommended to achieve an ethical responsibility of companies, and of their executives and owners, that would not be limited. Responsible executives and companies are the safeguards of the interests of the stakeholders, minority shareholders and the community. The leitmotiv should change from 'I am doing my best to diminish to a minimum my responsibilities' to 'I should behave responsibly toward my employees, all my shareholders, my country, my customers, ecology, and first of all toward my conscience.'

In the present state of affairs, there are too few whistle-blowers who have the courage to denounce overtly the crimes of companies against ecology or the stakeholders, to suffer the consequences, the ostracism of society, and the impossibility to find other jobs. An employee could agree to denounce his company in an extreme case, if there is a danger to the public or to the lives of people. But who would denounce overtly and without getting any remuneration a company that abuses the rights of minority shareholders? Let them solve their own problems; why should I risk my situation, my future, the bread-and-butter of my family, for some 'speculators' whom I do not know and who are attracted only by a quick profit on their investment in shares of my company? They would not have helped me in the same situation, so why should I help them? But if I would have something to gain from the publication of the information and if I do not risk anything, I could do it and also alleviate my conscience. The employees who would do it are only those who have a stronger allegiance to the community and to their conscience than to the company.

We could cite as precedents for the efficacy of denunciations, those that are made to the fiscal authorities and who come almost always from the close environment of the companies. If the IRS finds that it is ethical to encourage the denunciations, why should it not be encouraged also by the activist associations? But does the end justify the means, and can we remain ethical while encouraging denunciations, even of unethical acts? What is the alternative, let the majority shareholders or their companies wrong the minority shareholders? Is it not less ethical, is it a crime to denounce the criminals, or in the words of the Bible cited in this chapter 'The accomplice of a thief is his own enemy; He is put under oath and dare not testify.' There is a moral obligation to testify against a thief, unless you become his accomplice by not revealing his crime, even if you do not dare do so because you are afraid. Ultimately, if we do not find more efficient ways of safeguarding the rights of minority shareholders, we should envisage methods for denouncing unethical acts of companies and render them legitimate without any stigma, as it is probably the only way to resolve problems that could not be resolved otherwise, since crimes are performed usually in the dark.

The companies utilize extreme means to conduct their battles against their adversaries, even if they are dissident shareholders who dare oppose the executives and majority shareholders of their companies. They use the press, public relations agencies, investor relations firms, and even the Internet. But the press could also be used by minority shareholders in cases that could be of public interest. Unfortunately, the newspapers get tired of dealing with complicated cases, and in the long run they drop those cases for lack of public interest, or even as a result of heavy pressure of the companies that threaten to abolish their advertising budgets. An editor prefers a scandalous case of a rape over a tedious case of fraud of minority shareholders, who are often perceived as 'speculators'. But those minority shareholders can also employ public relations firms, which specialize in this domain, or organizations such as ADAM, which specialize in the protection of minority shareholders.

Another efficient method that could prevent the abuse of the rights of minority shareholders could be the distribution of rewards to the persons who divulge this wrongdoing of the companies, whether it is unethical or illegal. We enter here into a very problematic domain of the fidelity toward a company where we are employed, as the majority of the whistle-blowers would probably be employees of the companies concerned. Would the denunciations be anonymous like on the Internet? How could we distribute the rewards? And who will distribute them – the activist associations or another organization? Is it ethical to encourage the whistle-blowers? Would it be possible to employ this vehicle to get revenge from companies or executives who have not committed any fraud? How could we verify if the information is correct and make sure that the denunciations do not resemble precedent cases from totalitarian regimes?

We are educated since our childhood that it is prohibited to tell on your friends. The pejorative names for the telltales or tattletales are countless – whistle-blowers, stool pigeons, squealers, etc. Dante writes in the last verses of the Inferno, how the traitors and informers are punished in the lowest place of hell. Dante and Virgil enter Judecca, the lowest zone of Cocytus, where the souls of the traitors who betrayed their legitimate superiors and benefactors are totally immersed in the frozen waste. At the central and lowest point lies Satan, who devours Judas, Brutus and Cassius in his three mouths:

"That soul there, which has the worst punishment,
Is Judas Iscariot, my master said,
With his head inside, and kicking his legs.
Of the two others, who hang upside-down,
The one who hangs from the black face is Brutus;
See how he twists and says not a word;
And the other is Cassius, whose body looks so heavy."
(Dante, The Divine Comedy, Inferno XXXIV, 61-67, p.192-3)

It is incredible that out of all the criminals - those who have committed atrocious murders, genocides, rapes - the ones who receive the worst punishment are the traitors. It is not Pontius Pilate, who gave the order to crucify Jesus, it is not Julius Caesar who was an unscrupulous tyrant, it would not be Hitler if Dante would have lived in our times, but it would rather be Rommel, who 'betrayed' his fuhrer in order to save Germany.

Brutus and Cassius had to wait 1,600 years in order to be partially rehabilitated in the best historical play of Shakespeare 'Julius Caesar'.
(Brutus) "If then that friend demand why Brutus rose against Caesar,
This is my answer: Not that I loved Caesar less, but that I loved Rome more.
Had you rather Caesar were living, and die all slaves,
Than that Caesar were dead, to live all free men?
As Caesar loved me, I weep for him;
As he was fortunate, I rejoice at it;
As he was valiant, I honour him;
But, as he was ambitious, I slew him.
There is tears for his love; joy for his fortune;
Honour for his valour; and death for his ambition.
Who is here so base that would be a bondman?
If any, speak; for him I have offended."
(Shakespeare, Julius Caesar, Act III, Scene II, p.834)

The conviction that to denounce is an atrocious crime is inculcated in all peoples and religions. The Jews ostracized in the Diaspora the 'mousser', or the squealer, the person who denounced his brethren to the authorities, even if that brother was a thief or murderer. Everybody knows the awful fate of the squealers who denounce Mafia chiefs to the police. But the American and Italian police would have never succeeded in arresting Mafia leaders without the aid of the squealers of the Cosa Nostra.

Is it moral to denounce a crime committed by the Mafia to the police, in spite of the law of Omerta, which advocates a complete silence? Is it ethical to denounce an immoral act committed toward a customer or shareholder of a company by one of the company's employees? If he does not denounce his chiefs, the employee knows that truth will never be disclosed, and the company will continue to sell airplanes with damaged components, endangering the lives of the pilots, as was the case in many recent cases. Is the employee a squealer? If he believes in God and the Inferno, will he find himself in hell after his death in the vicinity of Judas and Brutus? If he is an agnostic, can he risk his career, the well-being of his family, the respect of his colleagues, in order to save the life of a pilot he does not know or to avoid the losses of a minority shareholder?

The employee will never denounce his superiors if society continues to treat him as a whistle-blower (pejorative connotation in the business world), a tattletale or sneak (pejorative connotation at school), an informer (pejorative connotation from the German Occupation), a stool pigeon (pejorative connotation in the Soviet Union), or a squealer (pejorative connotation from the criminal world). Maybe he would have the courage to denounce immoral acts, if he would be treated as a 'discloser', a neutral term meaning somebody discloses a fact, without a pejorative connotation. In this book the term whistle-blower is used, because otherwise the meaning would not be understood, but the meaning that the author of this book embraces is that of a discloser, and if it does not exist in the dictionary it is high time that it should be invented.

This discloser will not be ostracized but will be appreciated by the society in which he lives, as he will assist it to be cleaner and just. Many of the readers of this book will think of McCarthy who meant exactly the same thing when he urged intellectuals to denounce the 'communists' in order to have a cleaner society with no fear of the rising communism that endangered the existence of the free world. In most cases, nobody forced the people to denounce their friends, but those who did not cooperate did not get jobs and were ostracized.

What is therefore the difference between the proposals of this book and McCarthyism? McCarthy represented the authorities, he acted against the weak. Here is a completely opposite situation where the weak become organized against the powerful. It could be that in the future minority shareholders could become the strongest party, and activist associations would become too powerful. We have seen such inversions in the past in the Soviet Union, where the wronged proletariat became much worse and committed more atrocious crimes than the Tsarist regime that oppressed them. The author of this book believes in democracy and checks and balances, and hopes that the majority and minority shareholders will have a similar power without any one of them subjugating the other, exactly like the minorities are not subjugated nowadays in the United States like they were in the past, yet they do not subjugate the majority as well.

But we are aware that this argument will be raised, similarly to what the Jews in Russia called the 'wronged Kozak', meaning the Kozaks who organized pogroms against the Jews and pretended to be wronged by the persecuted Jews. Those who condemn Brutus, the rebel, the traitor, the squealer, to the pit of hell would have condemned as well the French Revolution which was against the legitimate power of the Bourbons, the American revolution which was against the legitimate power of the British, or the terrorists attacks of the Haganah, Etsel or Lehi in Palestine which were against the legitimate power of the British mandate. Those who condemn the whistle-blowers are in favor of the multitude of the immoral acts that are performed in companies against their stakeholders. The companies should be transparent ethically, without fearing anything from squealers, because when you have a clear conscience you do not need to be afraid to be discovered. Crime likes darkness, and the companies that do not conduct themselves ethically are looking for anonymity.

Moritatensanger:
"Und der Haifisch, der hat Zahne
Und die tragt er im Gesicht
Und Machheath, der hat ein Messer
Doch das Messer sieht man nicht.
Ach, es sind des Haifisch Flossen
Rot, wenn dieser Blut vergiesst.
Mackie Messer tragt 'nen Handschuh
Drauf man keine Untat liest.
An 'nem schonen blauen Sonntag
Liegt ein toter Mann am Strand
Und ein Mensch geht um die Ecke
Den man Mackie Messer nennt.
Und Schmul Meier bleibt verschwunden
Und so mancher reiche Mann
Und sein Geld hat Mackie Messer
Dem man nichts beweisen kann."
(Brecht, Die Dreigroschenoper, The Threepenny Opera,
Die Moritat von Mackie Messer, The Ballad of Mack the Knife, Act I, scene I)

"Streetsinger:
And the shark has teeth
And he wears them in his face
And Macheath, he has a knife,
But the knife one does not see.

Oh, the shark's fins appear
Red, when he spills blood.
Mack the Knife, he wears his gloves
On which his crimes leave not a trace.
On a nice, clear-skied Sunday
A dead man lies on the beach
And a man sneaks round the corner
Whom they all call Mack the Knife.
And Schmul Meier disappeared for good
And many a rich man.
And Mack the Knife has all his money,
Though you cannot prove a thing."

In order to denounce immoral crimes in companies, as for discovering the crimes of Mack the Knife, we have to be assisted by disclosers, as nobody sees the knives of immoral companies, which keep an impeccable facade and are assisted by the best lawyers and public relations. We need transparency otherwise nothing would ever be disclosed, and the law will never be able to safeguard the interests of the stakeholders, whether they are rich like Schmul Meier or poor like Smith. Therefore, only light can raise the curtain on the unethical acts of companies.

Moritatensinger:
"Denn die einen sind im Dunkeln
Und die andern sind im Licht.
Und man siehet die im Lichte
Die im Dunkeln sieht man nicht."
(Brecht, Die Dreigroschenoper,
Die Schluss-Strophen der Moritat, The Final Verses of the Moritat,
Act III, last scene)
"For the ones they are in darkness
And the others are in light.
And you see the ones in brightness
Those in darkness drop from sight."

Religious persons should conduct themselves morally as they believe that God examines their acts at every moment and nothing escapes him. For businessmen who are slightly less religious the fear of the disclosure of their acts to the public should replace the fear of God, because if they do not have anything to hide they will not have to fear anything. On the other hand if the employees utilize the liberty of disclosure to reveal the secrets of the companies to the competition or for reasons that have nothing to do with ethics, they would be subject to reprisals, exactly like the newspapers, which benefit from the liberty of the press and cannot disclose state secrets. The employees have to divulge only systematic and permanent cases of abuse which are inherent to the operations of the companies, which wrong the stakeholders, and which are backed by irrefutable documentation. They have to resort to outside bodies only after having exhausted all the internal bodies, which are meant to deal with those cases, such as the ethics officer, the superiors, the executives, the CEO, or even the board of directors.

There will always be cases where it will be argued that it is impossible to divulge a case as it is a state secret or a professional secret whose disclosure could endanger the company or the state. The most renowned case of a disclosure of a crime by act of conscience is probably the case of Colonel Picquart. One needs to have extreme courage in order to denounce his superiors, and bring against him the French army, the government and the majority of Frenchmen. But Picquart, imperturbable, testifies at the trial of Zola, after the latter wrote his famous 'J'accuse', where he accused the French authorities of concealing the truth about the innocence of Captain Dreyfus: "Pendant plus d'une heure, il expose, d'une voix tranquille, comment il a decouvert la trahison d'Esterhazy, les manoeuvres dont il a ete la victime et sa tristesse d'etre ecarte de l'armee. Les revisionnistes lui font une ovation. Apres quoi il est confronte avec ses anciens subordonnes, qui, tous partisans de Henry, l'accablent." (Troyat, Zola, p.274) "For more than an hour, he exposes, in a quiet voice, how he has discovered the treason of Esterhazy, the maneuvers that he was victim of and his sadness to be dismissed from the army. The revisionists make him an ovation. After that he is confronted with his old subordinates, whom, all colleagues of Henry, scorn him."

The modern history of business knows many similar glorious pages, where employees have denounced their companies at the risk of their career, their well-being and even their lives.

The transparency of companies will force every employee to ask himself at every moment the question: 'what is my ethical attitude toward this ethical problem?', because the following day his acts will be disclosed in the press or on the Internet, and his family, friends and congregation will learn about his acts. We will not have to ask ourselves anymore if our acts are legal or not, if they concur with the mission of the company and its ethical standards, but how they concur with our ethical standards, as we will not be able to hide anymore in anonymity. It will be like in the senate committees for the appointment of high officials, or with presidential candidates who are obliged to disclose their life transparently. Of course, we would have to beware not to resort to McCarthyism, to the open eye of the 'big brother', or to the denunciations of the sons and colleagues, as in the dictatorial regimes. The companies should be made transparent with measure and moderation and excesses will have to be condemned. Full disclosure should be made only on important cases, where the evidence is irrefutable, where there are no ulterior motives, and after having exhausted all other instance within the company.

The material advantages of the disclosers are often very high and outbalance the risks. In 1986, the US law, 'The False Claim Act' of 1863 was amended, and it encourages the disclosure of companies' fraudulent acts against the government. The discloser can receive up to 25 percent of the money that could be recuperated. The most renowned case is that of Chester Walsh and General Electric. In 1986, a manager of GE had conspired with an Israeli General to steal funds from the US military aid to Israel. The thieves succeeded in stealing at least $11 million, which was deposited in a Swiss bank account controlled by the Israeli General and the GE employee. Some employees of GE asked themselves how millions of dollars were transferred to a company that did not exist in the past. The control system of GE, the US army and the Israeli Army did not succeed in discovering the fraud. In 1992, GE admitted committing fraud and paid a sum of $69 million in fines. Twenty two GE employees were fired or punished. The discloser of the fraud was Chester Walsh, a GE marketing director in Israel, who succeeded during five years to gather documents, tape conversations and accumulate evidence of the fraud. Walsh and a non-profit organization sued the US government under the False Claims Act and received the sum of $11.5 million, which they shared.

This chapter is probably the most delicate chapter of the book, as it favors disclosure of immoral acts, which is contrary to our most innate hatred of whistle-blowers. It is after a long meditation that it was written, and following a conviction based on the analysis of case studies, the bibliography of this book, and a thorough empirical research. It is practically impossible to complete the ethical revolution without the publication of unethical acts of the companies. The measures envisaged will take a long time to be established and to prevail. In the short-term, it is principally the Internet and the disclosers which will be the vehicle for the promotion of ethics in business, as will be proved in the case studies in the empirical part of the book. Are the disclosers of those cases, Americans or Israelis, heroes? Or will they be condemned to join Brutus and Judas in hell? We shall prove their contribution to the transparency of the companies and the safeguarding of minority shareholders, and the readers of this book will be able to judge if they are traitors, martyrs or heroes.

Throughout the centuries, history repeats itself. Disclosers are called squealers and whistle-blowers by the legitimate forces that try to conceal their crimes. Progress is always linked with discoveries and disclosures, which the 'majority' tries to hide. Brutus makes a coup d'etat against a tyrant, although the majority worships Caesar. Galilei says 'e pur si muove' although the Church in 'majority' tries to silence him. The Dreyfusards try to acquit the poor Dreyfus although the 'majority' cannot admit that a Christian officer has betrayed his country. The financial tycoons of modern economy try to hide their actions, which transgress ethics and even the law. The only way to fight the prerogatives of the majority shareholders, to overcome the law of Omerta and to destroy the last bastion of totalitarian organizations, is to fling upon the windows of the companies and to render them transparent to all ethical critics. As the press safeguards the democratic regime; the Internet, the free access to information on companies, the possibility to reveal the cases which transgress ethics by the employees, should safeguard the interests of the stakeholders. The employees have to be the fiduciaries of the stakeholders and minority shareholders, like the quality managers are the fiduciaries of the customers. The Internet restitutes the Athenian democracy, as it is the modern Agora where nothing can be hidden. And when all companies will act openly, will be transparent, will not be able to hide dubious cases, the stakeholders of the companies, and especially the minority shareholders will have the possibility to be treated equitably.

6
ETHICAL FUNDS

"In a too limpid water, there are no fishes."
(Zen Proverb)

Before analyzing the advantages of Ethical Funds, which are self-evident, at least for the ethicists, one should also weigh their risks. This book, in spite of its clear-cut ideas, tries to raise contradicting opinions and weigh their merits. We shall see in the case of the French company how the majority shareholders justify their actions by the need to save the jobs of hundreds of employees, and how the minority shareholders are perceived as speculators who try to fish in filthy water. In other cases, the entrepreneurs maintain that it is necessary to risk other people's capital even at the risk of possible bankruptcy, in order to keep up with progress. The business world is blurred, 'dirty' and dangerous, and 'those who cannot suffer the heat should not stay in the kitchen'. On the other hand, the ethical world is perceived as a sterile world; theoretical and impractical for modern business.

Those arguments are very valuable, and majority shareholders may use as an example childbirth, which takes place in an environment of pain, blood and filthy water. But this example suits well those who maintain that business should not be dirty. Only in the twentieth century have we reached a high survival rate of newborns, as the sterile conditions in the hospitals help to overcome the dangers of infection in the birth process. In past centuries, where practically all births occurred at home, the mortality rate of newborns was very high. One should find therefore the proper measure between the natural conditions of the business world and the ethical prism that could sterilize them.

It is well known what happened in 1929, when there was no legal or ethical system to slow down the stock exchange speculations. It is not by sheer altruism that the SEC was established, following the recession and the collapse of the economic system in the '30s. The reform that ensued was opposed very strongly by the advocates of free enterprise. Thousands of people had to die from hunger, millions had to lose their jobs, a whole nation had to be impoverished, in order to convince the advocates of the 'dirty invisible hands', and that those hands had to be washed from time to time in order to obtain a minimal hygiene in business. In the year 2000 we have reached a similar crisis, and society should try to do its utmost in order to prevent the catastrophe that is about to be caused by the very dirty hands of some of the people who manipulate the economy.

The Ethical Funds will not choke the entrepreneurs, they will render life more difficult only to those majority shareholders who try to speculate to the detriment of minority shareholders, who use insider information, who cause the collapse of the shares' price in order to buy the minority shares at low prices, who act legally but unethically in order to sell artificially one subsidiary to another. All those cases will be treated at length in the empirical part of the book, but all of them could have been avoided if the companies would have submitted to the ethical screening of the Ethical Funds. In none of the cases did the minority shareholders put the companies' development in jeopardy, although they tried to oppose the unlimited greed of the majority shareholders, who often held much less than half of the shares.

The bibliography of business ethics is clearly divided between the optimists and the pessimists. We have those who are disgusted from the lack of ethics in the business world, the swindles, scams and schemes, and who hardly see a way to get out of this mud that prevailed in the past and will prevail forever. But on the other hand there are those who maintain their hope when they enter into the business world, who notice a favorable evolution, and who are convinced that they can change the negative trends and make an impact in their lifetime. Among those, we can find the businessmen and investors who have started the ethical funds movement, which has gathered tremendous momentum in the last years of the twentieth century.

Will there be any material change between the twentieth and the twenty first century? There is hope that the new ethical vehicles, such as the Internet, the activist associations and the ethical funds will make a substantial change in the negative environment that prevailed in the turn of the century.

The ethical funds were established in the United States, Canada and Great Britain, especially in the last ten years of the twentieth century. In 1999 in the US they had investments of more than two trillion dollars – 2,160 billion US$, 13 percent of all investments under professional management in the United States, invested in 175 funds. Those ethical funds succeed in maintaining better results than the average results in the stock exchange. The performances of the Domini 400 Social Index, comprising 400 ethical shares, beats regularly the S&P 500 and the Russell 1000, representing the average American securities. Europe is not so advanced in this domain as the US. Great Britain has already 34 ethical funds in 1999 with investments of 48 billion sterling pounds, the Netherlands have social responsible investments of more than one billion Euros, slightly more than Sweden, in Switzerland those investments amount to 0.8 billion Euros. The socially responsible funds in France have a much smaller scope – 0.4 billion Euros, in Germany about one quarter of a billion Euros, and in Israel they do not exist.

This book proposes to add as an investment criterion of the ethical funds the ethical relations between companies and minority shareholders. In this manner, investors in ethical funds, who are normally minority shareholders in the companies in which they invest, will be assured that their investment will be treated fairly. Ultimately, if most of the minority shareholders or the small investors will invest uniquely in ethical funds or in companies that behave ethically, we might be able to achieve the Lisistrata effect, when the strike of minority shareholders will force the companies to behave ethically, as without minority shareholders the companies will not be able to raise funds to operate their activities. The ethical funds could be a partial solution to the wrongdoing cases to minority shareholders that were analyzed in the case studies.

ETHICAL INVESTING

Ethical Investing, or socially-screened investing, is the placement of money in mutual funds, stocks, bonds, securities or other investments that are screened to reflect ethical, environmental, social, political or moral values.

Socially conscious investing is a way to build an investment portfolio that keeps pace with your conscience and reflects your beliefs, convictions and desire of change. It may involve avoiding companies with corporate practices you deem unacceptable or supporting acceptable ones.

Socially conscious investing grew from an early desire by many in the religious community to avoid investing in companies that profited from the sale of alcohol, tobacco or gambling products (sin and religious screens). The Pioneer Group, a group of 24 funds, has used a sin screen for almost seven decades, because the founder was a very religious man when he started the fund in 1928.

In the `60s, social investing grew even more popular as investors protested against the war in Vietnam. The Pax World Fund was started by Quakers and Methodists in the `70s to avoid investment in defense contractors in protest over the Vietnam War.

In 1972, the Dreyfus Corporation became the first traditional money-management house to add a socially screened fund, the Dreyfus Third Century Fund, that avoided investments in companies doing business in South

Africa, and sought out companies with good records for equal opportunity, safety, health and environmental care.

In the `80s, socially conscious investing entered the mainstream as our emotions were stirred by issues such as apartheid in South Africa, the environment and abortion. In 1982, the Calvert Group offered a fund with extensive social screens - the Calvert's Money Market Portfolio. By 1997, Calvert was offering a family of nine socially screened funds.

The really rapid growth in the number of funds having some type of social and/or environmental screening took place in the `90s. And the pace does not appear to be slowing. In the US alone, there are more than 100 funds with such screening. Ethical Funds are also very popular in Canada, Great Britain, and other countries in Europe.

There are a number of concerns generally shared by social investors. These include local community affairs, ecological and environmental issues, labor relations, minority and gender relations, military production, nuclear weapons and power, product quality and business practices.

Ethical investors, and the advisers who work for them, look at the social record of companies and investments on these issues to determine whether they are acceptable or desirable places to invest.

Researchers maintain that ethical investments can perform as well as conventional investments. In some cases they have performed better. For example, stock prices listed in the Domini 400, an index of 400 socially responsible US stocks, have grown 135 percent since the index was started in 1990 until the end of 1995. By comparison, the S&P 500 increased 120 percent.

The Domini Social Index (SRI) provides a broad market, common stock index for measuring the performance of portfolios with social constraints. The social investment research firm of Kinder, Lydenberg, Domini & Co. (KLD) constructs the DSI by identifying US stocks that pass a multitude of common social screens. The DSI was created on May 1, 1990 by KLD.

Two socially responsible indexes, the Domini 400 Social Index and the Citizens Index outperformed the Standard and Poors 500 Composite Index in 1999. Domini clocked in at 24.5 percent and the Citizens at 29.6 percent while the S&P gained 21 percent in calendar 1999. They have outperformed the S&P on a total return basis since their inception in 1990 and 1994 respectively. Nearly 70 percent of the largest socially responsible funds earned top ratings in 1999.

\# The most common criteria for investments in ethical funds are:

- Responsibility to the communities in which they operate, including the provision of products and services of long term benefit to the community, e.g. safety equipment. Charitable donations.

- A record of suitability, quality and safety of products and services.

- Adhering to environmental regulations and using technologies and products that are environmental friendly, non-polluting, conserving natural resources and energy, such as woodlands and forests.

- Progressive general approach to customers, suppliers and the public, as well as to industrial and employee relations - employment equity, welfare standards, labor safety practices, child labor laws.

- Operating within countries and regions that support racial equality, adhere to non-discriminatory hiring practices and avoid unreasonable exploitation of people generally.

- Deriving a majority of income from non-tobacco related products.

- Engaging in peace-based, non-military activities.

- Deriving income from activities that are non-nuclear and are not related to the production of nuclear fuel or waste.

\# Areas of Support of most of the Ethical Funds are: Education and training, Healthcare services and health and safety, Good employee relations, Equal Opportunities Policy, Policy statements audits and openness, Progressive community relationship and strategy, Effective corporate governance, Benefits to the environment, Energy conservation.

\# Other Areas of Support are: Multimedia and telecommunications, Mass transit systems, Pollution monitoring and control, Process control equipment, Recycling services, Renewable energy, Water management, Animals, Vegetarian foods, New textiles.

\# Areas of Avoidance of most of the Ethical Funds are: Alcohol, Gambling, Irresponsible marketing, Offensive advertising, Armaments, Oppressive regimes, Anti-Trades Union activity, Third World debt/exploitation, Pornography, Tobacco, Greenhouse gases, Mining, Nuclear power, Ozone layer depleters, Pesticides, Road Builders, Tropical hardwood, Water polluters, Animal testing, Fur, Meat/dairy production.

The basic structure for an ethical fund is as follows: Fund manager who decides what to invest in, Green/Ethical criteria - the stated guidelines and restrictions which the fund manager needs to be able to act, Share purchase when an investment is made, Ethical committee who monitors share purchases to ensure green/ethical criteria are being adhered to.

The Fund Managers ask companies to report in detail on their environment and social performance. They analyze the ethical information received by companies and conduct additional research from other sources to build up a complete picture in which a 'green' evaluation is made. They avoid companies deriving more than a negligible part of their turnover from oppressive regimes, or the arms, nuclear or tobacco companies.

Community Investing, supporting development initiatives in low-income communities both in the funds' countries and in developing countries, provides affordable housing, create jobs and helps responsible businesses get started. It is achieved mainly through Community Banks, Community Credit Unions, Community Loan Funds and Microenterprise lenders.

Social Venture Capital describes investing that integrates community and environmental concerns into professionally managed venture capital portfolios. The essence of venture capital lies between providing capital and management assistance to companies creating innovative solutions to social and environmental problems, and institutional investors investing on potential one billion dollar technologies.

The Funds aim to strike a balance between good and bad aspects of company activities, emphasize higher standards and positive aspects of corporate behavior, and influence companies to respond beyond the letter of the law to Ethical Criteria, through a system of 'constructive dialogue'.

Ethical Funds tend to avoid most of the largest corporations because they do not pass the ethical criteria. Smaller companies have more room for growth and can adapt more quickly to market changes, although of course they are more sensitive to market conditions.

Companies that perform according to ethical criteria ought to be more efficient, produce less waste, and have a more motivated and productive workforce, with less risk of prosecution, bad publicity, restrictive legislation, etc. The ethical fund managers who are active in their research will know a lot more about the companies they invest in and therefore can make more informed investment decisions.

Most of the Ethical Funds invest in companies that adhere to the Ceres Principles and publicly affirm their belief on their responsibility for the

environment in a manner that protects the Earth. The Principles are - Protection of the biosphere, Sustainable use of natural resources, Reduction and disposal of wastes, Energy conservation, Risk reduction, Safe products and services, Environmental restoration, Informing the public, Management commitment, Audits and reports, Disclaimer.

The Ethical Funds' mission is to be the providers of Socially Responsible Investments (SRI) products and services. In order to achieve this mission, the funds should have a clearly defined business culture and ethical commitment, that consists of the following principles:

- Maintaining the highest ethical standards when dealing with their clients. Placing the clients' interests first, reflecting the social concerns of the clients in their recommendations, and fully disclosing their means of compensation.

- Providing clients with the broadest possible range and highest quality of investment choices, social research, and service options.

- Operating in a fiscally responsible manner. Generating an adequate profit margin, while reinvesting in the continued growth of the funds.

- Encouraging, developing and maintaining a high level of professional standards and education, mainly in Socially Responsible Investing.

- Supporting public education to promote the relationship between financial decisions and the public good. Supporting social activism by providing information on shareholder activism and boycotts.

- Operating in a socially responsible manner. Balancing the interests of all their stakeholders - clients, staff, products suppliers, research companies and local communities.

ETHICAL FUNDS IN THE US

There are more than 175 US Ethical Funds, very diverse in nature. In 1982, there were only about 20 funds in the US. They grew to about 30 in 1991, 55 in 1994, 139 in 1997. Most of them are focused on environmental and social screening, and invest either in fixed income securities or growth shares.

$2.16 trillion is invested today in the US in a socially responsible manner, up a strong 82 percent from 1997 levels, according to a study released on November 4, 1999 by the nonprofit Social Investment Forum. It accounts for

roughly 13 percent of the $16.3 trillion under professional management in the US.

Since 1997, total assets under management in screened portfolios for socially concerned investors rose 183 percent, from $529 billion to $1.49 trillion.

96 percent of the US socially screened portfolios avoid tobacco, 86 percent gambling, 83 percent alcohol, 81 percent do not invest in weapon industries. 79 percent have an environmental screen, 43 percent human rights, 38 percent labor issues, 23 percent birth control/abortion, and 15 percent animal welfare.

Over 120 institutions and mutual fund families have leveraged assets valued at $922 billion in the form of shareholder resolutions. These institutional investors used the power of their ownership positions in corporate America to sponsor or co-sponsor proxy resolutions on social issues.

The fastest growing component of socially responsible investing is the growth of portfolios that employ both screening and shareholder advocacy, with assets growing 215 percent from $84 billion in 1997 to $265 billion in 1999.

Funds that are mainly environmental are for example - AFW, Alliance Global, Better Than Bonds, Global (2 funds), Green Century (2), INVESCO (3), Kemper. Funds that are mainly social responsible, with social and sin screens, are - Ariel (3 funds), Bridgeway, Calvert Group (9), Citizens Trust (7), College Retirement Equity Fund, Common Sense Trust (10), Delaware Quantum, Dreyfus 3rd Century, Lincoln Life, MFS (2), Neuberger & Berman, NWQ (3), Pax World and Growth Funds, Pioneer (24), Rightime, Security Benefit, Smith Barney, Social Responsibility, Vermont National Bank.

Funds that have religious screens are - Amana (Islamic, 2 funds), American Trust Allegiance (Christian Science), The Aquinas (Catholic, 4), Catholic Values Investment, Islamica (2), Lutheran Brotherhood (7), MMA (Mennonite, 3) Noah (Judeo-Christian), The Timothy Plan (Christian).

Funds for affordable housing and community development are - Alternatives Federal Credit Union, Community Capital Bank, Self Help Credit Union, South Shore Bank. American Mutual Fund and Washington Mutual Investors Fund do not invest in companies with revenues from alcohol or tobacco. Domini Social Equity Fund is based on the Domini Social Index, Meyers Pride Value Fund invests in companies with progressive policies toward gays and lesbians.

Other funds - Beacon Cruelty-Free Value, Delaware Quantum, DEVCAP Shared Return, Eclypse Ultra Short Term, Hudson Investment, Laidlaw Covenant, New Alternative, Parnassus (4), Stein Roe Young Investors, Total Return Utilities, Victory Lakefront, with African-American/diversity screen, Wasatch Funds (5), with a Latter Day Saint screen, Women Pro-Conscious Equity Mutual.

First Affirmative Financial Network (FAFN) is the original nationwide network of financial advisors specializing in Socially Responsible Investments (SRI). FAFN is an active member of the Social Investment Forum. Co-op America, a nonprofit organization, began working together in 1989 with FAFN to convey the mission to promote social and environmental change within the marketplace and their 55,000 members.

The Interfaith Center on Corporate Responsibility (ICCR) is a 26-year old coalition of 275 Protestant, Roman Catholic and Jewish institutional investors with combined portfolios of over $75 billion. The ICCR coordinates the ethical and corporate responsibility programs of its members, utilizes dialogue with corporate management and conducts aggressive public campaigns challenging corporate irresponsibility.

ETHICAL FUNDS IN CANADA

In Canada there are 15 mutual funds that screen the companies in which they invest according to social or environmental criteria. All of these funds invest broadly in the sectors of the economy to achieve a diversified portfolio. The 15 funds are operated by four mutual funds companies.

The 1997 Canadian Ethical Money Guide by Eugene Ellmen (Lorimer, December 1996) takes the investors through the challenges of managing money in a socially-responsible way. It rates all the Canadian ethical funds, recommending the ones with strong economic and social mandates.

Desjardins Environment Fund. Sponsored by Desjardins Trust, a subsidiary of the Quebec-based Desjardins system of caisses populaires, this fund invests in environmentally conscious Canadian corporations.

Investors Summa Fund. Operated by Investors Group, the largest mutual company in Canada, the Summa Fund screens for alcohol, tobacco, gambling, military weapons, pornography, environmental policies and repressive regimes.

Dynamic Global Millennia Fund, managed by a Toronto-based mutual fund group, invests in environmentally screened companies.

Clean Environment Funds. A Toronto-based mutual fund company operating four funds investing in companies reflecting the concept of sustainable development. These companies include, but are not limited to, waste cleanup firms and environmental companies.

Ethical Funds. Owned and controlled by the Canadian credit union system. The eight funds operated screens for industrial relations, racial equality, tobacco, military production, nuclear energy and environment.

Ethical Funds are among the top-performing mutual funds in Canada, delivering impressive returns to over 110,000 investors from nearly 1000 credit union branches across Canada.

Ethical Growth Fund performed in the top quartile (the top 25 percent of all Canadian relevant funds) for every time period measured - 1,2,3,5,10 years and ranked third out of 84 Canadian Equity Funds for performance over the past 10 years.

Ethical North American Equity Fund was fourth out of the US equity funds over one and three years, and ranked tenth out of a total of 1,331 Canadian mutual funds for past year performance. Most of the other Ethical Funds performed in the top quartile over one, three and five years.

ETHICAL FUNDS IN THE UK

Great Britain had in 1999 34 ethical funds with investments of 48 billion sterling pounds.

The most prominent Ethical Fund in the UK is Friends Provident Stewardship Fund, with Funds of £570M, launched in 1984. Other funds are - Credit Suisse Fellowship Trust, Framlington Health Fund, Abbey Life Ethical Trust, Allchurches Like Ethical Trust, Jupiter Ecology Fund, Scottish Equitable Ethical Unit Trust, Sovereign Ethical Fund, Eagle Star Environmental Opportunities Trust, TSB Environmental Investor Fund.

Other Ethical Funds in the UK are: D.J. Bromidge Ethical Investment Fund, Barchester Best of Green Fund, Citibank Life Green Fund, Ethical Investors Group Cruelty Free Fund, Genesis Fund, Lincoln National Green Fund, MI Environment Fund, Skandia Third World Fund, Commercial Union Environmental Trust, Clerical Medical Evergreen Fund, CIS Environ Trust,

NPI Global Care Fund, Skanida Ethical Selection, Cooperative Bank Ethical Unit Trust, Equitable Ethical Trust, HTR Ethical Fund.

The prominent specialist broker funds in the UK are: GIFA Managed Unit Trust portfolio, Minerva Managed Unit Trust portfolio, Barchester Best of Green Offshore Fund. Albert E. Sharp Ethical PEP offers a direct equity portfolio for PEP investment. Three ethically screened banks - Co-Op ICOF, Triodos Bank, Shared Interest, offer various accounts and loans to acceptable applicants. Ecology Building Society is a green building firm.

Friends Provident was founded by Quakers in 1832, and has become one of the most progressive and successful insurance and investment groups in the UK. With more than two million individual policies and accounts, and subsidiaries and affiliates across 14 countries, the Friends Provident group manages assets of about £17 billion worldwide.

Who is 'Ethics Man'? 44 percent of Friends Provident Stewardship unitholders are women, a much higher proportion than average, 48 percent are over 54, and only 10 percent are under 35. 61 percent are in professional or managerial occupations, compared to less than 20 percent of the total adult population.

Stewardship unitholders belong to various organizations - The National Trust (42 percent), Amnesty International (28 percent), Greenpeace (25 percent), Friends of the Earth (24 percent). 19 percent are Anglicans, 11 percent are Quakers, 7 percent are Methodists. 90 percent give money to charity, 53 percent do voluntary work, 90 percent recycle household waste, 57 percent avoid unnecessary car journeys.

The UK Social Investment Forum's primary purpose is to promote and encourage the development and positive impact of Socially Responsible Investment (SRI) throughout the UK. SRI has been defined as an investment that combines investors' financial objectives with their commitment to social concerns such as social justice, economic development, peace or a healthy environment. The forum was launched in 1991. The main objectives of the forum are - to promote the understanding of socially responsible investment and support a greater sense of social accountability among investors. The forum initiates and publishes research for required changes in legislation and company policies.

ETHICAL FUNDS IN FRANCE

"One of the sources of the renewal is the creation in 1997 of the Arese (Analyses et recherches sociales sur les entreprises) by Genevieve Ferone, that is examining the social conduct of Michelin after they announced simultaneously a plan to lay-off 7,500 employees in Europe and an increase of 17 percent in their semi annual profits. Offshoot of the Caisse des depots et des Caisses d'Epargne, the Arese rates the companies quoted in the stock exchange in function to their attitude toward their customers, suppliers, subcontractors, shareholders and the society. Its objective is to promote the socially responsible investment (SRI) which integrates social and/or environmental criteria in every investment decision, without overlooking the pursuit of financial profitability. Social relations, training, work conditions, employment and remuneration policy, methods of participation of employees, are examined very thoroughly. 'Those criteria supplement the traditional financial analysis while enabling the creation of investment funds which are specific for particular customers or institutions', says Genevieve Ferone.

'The researches of the Arese are a precious tool. On top of his private beliefs, everybody needs precise landmarks', believes Marc Favard, who manages the Ethical Funds at Meeschaert Roussel, one of the pioneers of the ethical investment in France, with sister Nicolle Reille, who has created the funds Nouvelle Strategie 50 in 1983, in order to invest the funds of the religious congregations while respecting morals. Later on appeared the sharing funds, or solidarity funds. There are about 30 of them, with rather low investments between 50 to 100 million francs, most often tied up to housing or to loans to very small firms. Between their promoters, many syndicalists or associative militants, coming from the banking world, like Maurice Bideau, the president of NEF (Nouvelle Economie Fraternelle). 'The saying is that you lend only to rich people. The reality is on the contrary that the small projects are easier to manage' maintains Maurice Bideau.

The Guide of Ethical Investment, published by the weekly la Vie and the monthly Alternatives economiques, specifies an impressive list of solidarity investments for all tastes, with a precise cartography. All of a sudden we find that the ethical funds address a large public, much larger than the solidarity funds, with a much tighter efficiency constraint. They do not represent much, but are becoming influential in the financial markets in rapid growth, where the imperative of short term return on investment shows its limits... The economical risk has on the contrary to be evaluated permanently by the minority shareholders, as has shown the case of Eurotunnel: the investment of the small citizen and the high return on investment promised to the small investors (70 percent of the equity) has not subsisted to the overindebtedness of the company. 'The pension funds are aware also of the social or environmental risk incurred by companies without sufficient standards. They include social criteria in the good corporate governance', reminds Genevieve Ferone, who worked for five years at a US law firm on large American

pension funds, and wrote a book on the subject - 'Le Systeme de retraite americain, les fonds de pension'. Genevieve Ferone. PUF." (Thiery Nicolas, A la decouverte des fonds ethiques, The discovery of ethical funds, La Tribune, 19.10.99)

The principal ethical funds of the new generation in France in 1999 are:

- Actions Ethique de Meeschaert (FCP, 30MF in recent assets, founded in July 1998)

- La Sicav EuroSocietale d'ABF Capital Management (recent assets of 85MF, founded in May 1999 for the Euro zone)

- Macif Croissance Durable (FCP, recent assets 65MF, founded in May 1999)

- France Expansion Durable d'Expertise Asset Management (Sicav, recent assets of 65MF, founded in July 1999) The manager is Isabelle Delattre who chooses securities from the index SBF 120, with some securities of the SBF 250, when a share interests her especially. She then requests punctually a study from the Arese. 54 percent of the invested portfolio is today in securities of the CAC 40.

- Ecureuil 1,2,3 Futur des Caisses d'Epargne (Sicav, created in October 1999) For Erik Pointillart, director of development of Ecureuil Gestion, 'the investment should reach 500 millions francs, and then one billion in cruise speed.'

- RG Hommes, Terre, Expansion of Robeco. With its experience in ethical funds in the Netherlands, where the company manages 1.2 billion francs, Robeco will launch in France by the end of October a FCP of international shares based on social ethical, environmental and economical criteria." (Lambert Agnes, Les fonds ethiques s'ouvrent aux particuliers, Ethical funds open to the public, La Tribune, 24.9.99)

ADDITIONAL DATA

"Here are some findings from a 1994 survey conducted jointly by Cone and Roper: Seventy-eight percent of adults said that they were more likely to buy a product associated with a cause they care about.
. Sixty-six percent of adults said that they'd be likely to switch brands to support a cause they care about.
. Fifty-four percent of adults said they'd pay more for a product that supports a cause they care about.

. After price and quality, 33 percent of Americans consider a company's responsible business practices the most important factor in deciding whether or not to buy a brand.

The number of people who want to 'vote with their wallets' is growing toward critical mass. There's a paradigm shift occurring. As one indicator, the Council on Economic Priorities' handbook Shopping for a Better World, which rates the products available in supermarkets based on their degree of social responsibility, has sold over 1 million copies since 1991." (Cohen and Greenfield, Ben & Jerry's Double-Dip, p. 48-9)

The social-performance report of Ben & Jerry's covers: social activism, customers and their needs, environmental awareness, supplier relationships, use of financial resources, financial support for communities, quality of work life. "The social audit is useful to shareholders or prospective shareholders. Theoretically those folks are reading the company's financials. If a shareholder is interested in investing her money based on social criteria as well, she should be able to read the company's social audit and make her investment decision based on being able to compare different investment opportunities. More and more companies are doing social audits in one form or another. The Body Shop and Whole Foods Markets do social-performance reports similar to what Ben & Jerry's does. Other companies publish environmental-impact disclosures and statements of social responsibilities: Patagonia, Reebok, British Airways, Volvo, Philips Electronics, Sony, Compaq, Intel, and IBM, among others." (Cohen and Greenfield, Ben & Jerry's Double-Dip, p. 251)

The firm Kinder, Lydenberg, Domini, established in Boston proposes since 1990 the Domini 400 Social Index, composed of 400 socially responsible American securities. It excludes the fields of tobacco, alcohol, gambling, nuclear, and companies with more than two percent of its turnover in military contracts. In total, half of the securities of the index S&P 500 enter in the composition of the Domini 400. The results are indisputably in favor of the ethical funds: by the end of July 1999, the Domini 400 index increases by 24.08 percent annually against 20.27 percent for the S&P 500. Over five years, the annualized performance amounts to 28.04 percent for the ethical index, which is two points higher than the larger index of Wall Street. Many American companies propose ethical funds to their employees in their pension funds. This has brought up a favorable and durable dynamism for the development of those products.

7
ACTIVIST ASSOCIATIONS, 'TRANSPARENCY INTERNATIONAL', 'ADAM'

"Dr. Stockman: And just look here, Katherine – they have torn a great rent in my black trousers too!
Mrs. Stockman: Oh, dear! – and they are the best pair you have got!
Dr. Stockman: You should never wear your best trousers when you go out to fight for freedom and truth."
(Ibsen, An Enemy of the People, Act V)

The rights of minority shareholders are tightly linked to the evolution of the rights of stakeholders. The companies are controlled today in most cases by majority shareholders who own often less than 50 percent of the shares but who manage to control the boards of directors. From the moment that the stakeholders will be represented in the boards of directors, the rights of minority shareholders will also be safeguarded. The majority shareholders justify their absolute control of the company by the fact that they have invested their capital into the company. Nevertheless, Estes and many other authors maintain that the stakeholders invest also in the company, often much more than the majority shareholders.

"But the corporation has other constituents as well: the workers, customers, suppliers, community, and the greater society. These other stakeholders are investors too, and they often risk far more than financial investors. Employees invest in the corporation. They bring their education, skills and experience – often gained at substantial personal expense – to the job. They invest time, energy, and too often their health. They invest their careers, careers that can be effectively wiped out in a casual layoff or relocation decision... Customers invest in the corporation. Their monetary investments are often greater than those of stockholders... Like workers, suppliers are investors too. They may commit production facilities, install special equipment, redesign products, and provide financing to their corporate customers. They have a right to expect fair treatment and a fair return on their investment.

Communities – neighbors, towns, cities, counties, and states – invest in corporations. They provide much of the infrastructure, such as streets and bridges, water and sewer systems, and police and fire protection, without which the corporation could hardly function.... Communities are investors and deserve a fair return on investment as much as stockholders. The nation –

society – invests in the corporation. It provides the social capital and structure, without which we would face the brutal anarchy of the cave dweller. Our society supports the democratic system that allows the corporation, and the rest of us, freedom of movement and action. It provides protection for the free enterprise system.

Nations also grant specific benefits to corporations, such as investment incentives, tariff protection, research subsidies, defense contracts, and tax benefits including investment tax credits, accelerated depreciation, and foreign tax credits. Employees, customers, suppliers, communities, and society are all investors, but the corporation is not accountable to them. It reports regularly and comprehensively to stockholders, almost never to other stakeholders." (Estes, Tyranny of the Bottom Line, p. 4-6)

If you analyze which funds effectively finance the company, we shall notice in most of the cases that the funds of the shareholders who control the company contribute only a minimal part of the necessary funds for the functioning of the company. In many cases those who control the company are the executives who have not invested anything in the company even if they own its stocks. In the cases of the founders, they have invested in the initial phases of the company or when the shares' prices were not so high, and those who have invested the largest sums in equity are the minority shareholders who not only are not represented in the boards of directors but also have invested when the shares' prices were very high, mainly at public offerings.

Furthermore, the original investors of the company have often sold their shares on the stock exchange, and the new shareholders have not invested into the company but paid to the other shareholders for their shares. Thus, the company has not profited from the appreciation of the price of the shares, especially if it does not issue new shares. The suppliers, willingly or not, finance the company which utilizes their credit to finance the working capital. The clients finance undoubtedly the company, as it is their revenues that generate the profits of the companies. The creditors finance the company, as their financial leverage finances sometimes two or three times more than the equity. It is superfluous to state that the financing of the community and the state is so high, that in some stages, especially in the first ones it can amount to a third or even more of the total financing.

In the last years, we witness in the US, and to some extent also in France, a growing social activism of the shareholders and in many cases they succeed in changing the decisions taken by large companies in the US: "The world-wide phenomenon observed as a growth in shareholder awareness comes under the general term of 'government of companies' or corporate governance. This phenomenon involves an increased interest in two categories of concerns linked to the internationalization of the capital of large industrial and financial conglomerates. The first category of concerns, already well recognized in France, regards questions directly relating to the rights of shareholders: company policy on information, distribution of profits, the organization of the board of directors, remuneration and protection of managers, etc.

The second category, not yet well known in our country but more widely discussed on the other side of the Atlantic, covers questions related to the general direction taken by management in response to a movement which could be termed 'social activism of shareholders'... Numerous recent initiatives by shareholders in the United States - 'General Electric sells its aerospace division to Martin Marietta under pressure from the Sisters of Notre Dame de Lorette; - The sisters of the Charity of the Holy World force Kimberley-Clark to sell its tobacco division; - The Lourdes Medical Centre forces the management of Pfizer to change their strategy; - The Sisters of Sainte Catherine de Sienne win a lawsuit against Wal Mart..." (Richardson, World Ethics Report, Leroy, Development of Social Activism amongst Shareholders, p. 161)

In shareholders' meetings in the US there are hundreds of resolutions that are adopted every year as a result of the activism of the shareholders, who are mainly minority shareholders. The most dominant organizations in their activism are religious associations, proactive associations of shareholders, often with women dominance. "The spiritual heart of this movement is a New York non-profit organization, the Interfaith Centre for Corporate Responsibility... For the last twenty-five years, ICCR has organized a coalition of 275 institutional investors, Protestant, Jewish and Catholic, who together represent a share portfolio with a total of value of 45 billion dollars. This organization co-ordinates the activity and voting of its members at shareholder meetings. Each year, it also publishes the astonishing growth of external proposals put forward by shareholders at general meetings of American publicly-owned companies...

In the United States, ownership of shares is popular and the American financial system is favorably disposed to direct intervention by shareholders in the business affairs of a company. Contributory pension funds are managed by organizations without links to the banking system and they are also subject to managements by vote. In addition, the invested capital allowing a shareholder to propose a motion at a company general meeting is low, being only one thousand dollars. To be included in the agenda of a general meeting, any resolution must also be recorded by the company; and the minutes are controlled by the American Securities and Exchange Commission." (same, p. 162)

The situation in France is much less evolved. The COB has exhorted since 1990 the shareholders to exert a more active role in their voting rights. Shareholders' associations, as the ANAF, managed by Marcel Tixier and the ADAM managed by Colette Neuville, have been mobilized to safeguard the rights of the shareholders. The shareholders' meetings are much more animated and the approach of the minority shareholders is much more critical. Unfortunately, the rights of the minority shareholders are very limited, as we shall notice in the case of the French company in the empirical part of this book. "It has in fact been calculated that it is necessary to have at least 300 million francs worth of shares to propose an external resolution at an annual general meeting of one of the first twenty companies listed on the Paris stock exchange. To this barrier must also be added the numerous statutory limitations on the minority shareholder's right to vote through such ploys as double voting rights and other restrictions which the President of the National Association of French Shareholders does not hesitate to term as 'vote-rigging'." (same, p. 163)

In France, the members of the boards of directors of the largest companies are often the same persons and are elected in most of the cases by the CEOs which they are supposed to control. "The independent directors exercise a sort of counter-balancing power, but to whose benefit? To the benefit of all the shareholders, states the Vienot report; but the COB (commission operating the French Stock Exchange) is more precise. The mission of the independent director should essentially be to protect the minority shareholders of a company. Another pillar of corporate governance is the establishment of management committees with specific functions.

The Vienot report recommends the creation of a committee for the appointment of directors as well as an external audit committee having as its main aim the confirmation of the company and group accounts. These committees, frequently found in Anglo-Saxon countries and already in certain companies in France, to a certain degree respond to minority shareholder expectations of transparency. With the same aim of protecting minority shareholders, the COB concluded the Vienot report by supporting attestations of equity and the call for a generalization of opinion from the board of the targeted company when a financial operation is complex, when there is a simultaneous sale of assets or there are contradictory interests at stake." (Richardson, World Ethics Report, Endreo, Protection of Minority Shareholders in France, p. 186)

The minority shareholders are not conscious of their power, in the same way that the people were not conscious of their strength before Rousseau, Voltaire and the French revolution. A large number of minority shareholders act like Candide and are convinced that everything is for the best in the best of the world, and that they should continue to lose in the long run like the gamblers who lose at the casino. There are very few militant minority shareholders and very few organizations that safeguard their interests like ADAM, managed by Mme. Neuville in France. The power of these individuals and organizations is very limited and if they sue the companies they often lose. But they ignore that they possess the absolute power, the Armageddon weapon, the absolute weapon, and if they use it they could collapse the Philistines' temple. But Samson, who is blind and thinks that he has no power, does not have to die with his persecutors. The minority shareholders can cease to invest in companies that do not behave ethically and in parallel invest uniquely in ethical funds. They could also, if they do not want to incur any risk, invest their money in savings deposits and be satisfied with 5 percent interests per annum.

Majority shareholders and the companies cannot operate without minority shareholders, as the majority shareholders invest effectively in most of the cases only about 30 percent of the equity in order to obtain control of the company and the remainder is invested by the minority shareholders, who own in fact the majority of the shares without having any control of the company. In paraphrasing a well-known 250-year-old maxim, the minority shareholders should say – no investment without representation! Furthermore, the majority shareholders do not lose in most cases from their investment, as they know when to sell and buy the shares with their insider information.

Ultimately, the minority shareholders invest effectively almost all the capital in absolute terms, not in number of shares, as they invest at the highest prices at the offerings, and the majority shareholders manage to recoup all their investments while selling part of their shares at offerings or in the market at high prices, thus risking henceforward only part of their return on investment but not of their capital that they have recouped. In many cases the majority shareholders profit from a collapse of the price of the shares and buy from the panicked minority shareholders shares at 10 percent of their previous prices, thus increasing even more their ownership and their profits.

When the situation stabilizes and the prices of the shares increase again, they sell once more their shares at the higher prices to the new minority shareholders, and like in a perpetuum mobile, they always increase their ownership and profits to the detriment of the minority shareholders. This circus continues invariably for more than 100 years, as the same norms that prevailed in the times of the Second French Empire and the robber barons still exist in the year 2000. Suckers never die, they are just replaced, and as nobody warns them, least of all the SEC, the rich get richer and the poor get poorer in the stock exchange, and the more it changes the more it remains the same.

It is stunning how democracy has evolved dramatically in the last few years, but how democracy in business has remained retrograde. Heraclites has said that cattle is driven to the water with a stick, and the same law prevails probably with humankind. There needs to be a catastrophe in order to instigate drastic change, as only after World War II did the world reach the conclusion that the best regime is the democratic regime, and the communist economies needed to collapse in order to change their totalitarian regimes.

The minority shareholders have probably not suffered enough, as the French people before the revolution of 1789, or the American people before the War of Independence in 1766. At the end of the twentieth century the stock exchange has reached new records; many minority shareholders got rich by investing in high-tech companies, and the scandals of the '80s are long forgotten. In searching in the world bibliography for the subject of this book we discover that almost nothing was written previously on this matter, probably because it does not interest the minority shareholders. Do we need a worse catastrophe than in 1929 in order to convince the minority shareholders to take their fate in their hands and exert the power that they can possess?

The author of this book is not so pessimistic and is convinced that even without a catastrophe evolution is inevitable and in five or ten years at most there will be a drastic change in ethics in the relations between companies and minority shareholders. We need to publish theses, books, articles on this subject, we need to introduce new norms, we need to use the Internet and other vehicles to augment the democracy of companies and assist the minority shareholders. We have to remember that there has never been a revolution in the US to abolish racist laws, there has never been a revolution in South Africa to abolish apartheid, and there has never been a revolution in the Soviet Union and its satellites to establish capitalistic democracy.

The dictatorial regimes of Spain, Portugal, Argentina, Chile or Greece have disappeared almost without bloodshed, although they were established in civil wars and bloody revolutions. The reason for this evolution without revolution was that the dictatorial regimes were ostracized and boycotted by the democratic countries, which have also ostracized the regimes of the Soviet Union and South Africa. In the same manner we need to ostracize and boycott the companies that will not conduct themselves ethically in general and toward the minority shareholders in particular.

Minority shareholders are waiting for their leaders, their Martin Luther King, their Nelson Mandela or their Ben Gurion. They are waiting for their 'Altneuland', their 'Contrat Social' or their 'Kapital'. Business ethics is not merely a nouvelle vague, a new wave, an ephemeral fashion, a gimmick, a buzz word. This is the new level of evolution of business, after the taylorism, the marketing, the organization, the quality and the ecology. The time of business ethics has arrived and it will remain forever. But the victory of ethics cannot be achieved without achieving also ethics toward the minority shareholders.

We should however be careful not to succumb to the tendency to pay artificial tribute to ethics as many companies are doing today, by having Codes of Ethics and not practicing them. As it is not politically correct to express oneself with pejorative terms toward women, Afro-Americans or Jews, the majority of businessmen declares its profound allegiance to business ethics but continue to act as in the past in their intimate circles. Eventually, they could hire an ethics officer, ethics consultants, or finance an ethics cathedra, to use them as Adam's leaves to cover the moral nudity of their companies.

We know how the 'robber barons' have alleviated their conscience by donating millions of dollars to build museums, universities or hospitals. According to their ethical norms and the norms of their followers to our days, they can despoil the rights of minority shareholders, cheat their customers and suppliers, destroy the ecology of entire nations, and make amends for it by giving to society a small percentage of what they robbed and usurped. And society, in order to thank them, nominate them as doctors honoris causa, give them the legion of honor, or the award for the best industrialist or exporter.

The only way to act against those ethics criminals is by organizing a campaign led by the activist associations that will ostracize the unethical businessmen instead of envying them, to refuse their donations, to nominate them doctors deshonoris causa and to put them on the black list. For them, appearances are very important and they invest considerable amounts in public relations in order to save face. We should only change their rules of the game, as those who should lead in the business world should be the ethical businessmen. It would be like being members of an exclusive club, where the ethics criminals would not be admitted, even if they try to redeem themselves. This subject of recognition has been treated so far and will be treated further on in many angles, as it is crucial for making ethics prevail in business. The case studies will illustrate those aspects in a very vivid way as the present tendency is completely opposite, since in many cases society ostracizes the disclosers or whistle-blowers and venerates the unethical businessmen.

We have already mentioned the activist minority shareholders, but we should emphasize also the worker-owners, as a vehicle to safeguard the rights of minority shareholders. The activist minority shareholders were already responsible for the significant improvement of competitiveness and financial results of many American companies in the last ten years of the century. Companies such as General Motors, IBM, Eastman Kodak, Westinghouse, and Sears Roebuck have improved their performance as a result of an intervention of activist shareholders. The 100 million salaried in the US possess through their pension and other funds the majority of shares of a large number of companies. "There are now over 10,000 American ESOPs, including huge companies such as United Airlines, Avis Rent-a Car, and Weirton Steel, and there is evidence that they are more responsive to their employees and their customers. Studies show that worker-owners are more productive and deliver higher quality, with Avis now number one in ratings of customer satisfaction.

Hundreds of ESOPs and cooperatives, including large worker-owned factories, practice sophisticated forms of workplace democracy. They are proving effective in job creation and retention, and are responsible for saving hundreds of jobs during the epidemic of factory closings in the last decade. According to polls, including one by Peter Hart, economic democracy makes sense to most Americans; approximately 70 percent say that they would welcome the opportunity to work in an employee-owned company. Employee ownership in the United States has grown fifty-fold since 1974, with employees being the largest shareholders in more than 15 percent of all public companies.

The cutting edge is in the Fortune 500, where by 1990 the percentage of employee ownership was 11.7 percent in Ford, 9.3 percent in Exxon, 10 percent in Texaco, 16 percent in Chevron, 24.5 percent in Procter & Gamble, 18.9 percent in Lockheed, and 14.5 percent in Anheuser-Busch. By 1995, employee ownership was higher than 30 percent in huge companies such as Kroger, McDonnell Douglas, Bethlehem Steel, Rockwell International, Hallmark Cards, Trans World Airlines, U.S. Sugar, and Tandy Corporation. Thirteen percent of the labor force – 11 million workers – are employee owners, more than the number of private sector union members. The total value of stock owned by workers in their own companies now exceeds $100 billion." (Derber, The Wilding of America, p. 158-9)

TRANSPARENCY INTERNATIONAL

Transparency International is a non-governmental organization, operating in about 90 countries and dedicated to increasing government accountability and curbing both international and national corruption. The movement has multiple concerns:
humanitarian, as corruption undermines and distorts development and leads to increasing levels of human rights abuse.
democratic, as corruption undermines democracies and in particular the achievements of many developing countries and countries in transition.
ethical, as corruption undermines a society's integrity.

practical, as corruption distorts the operations of markets and deprives ordinary people of the benefits that should flow from them.

Combatting corruption sustainably is only possible with the involvement of stakeholders, which include the state, civil society and the private sector. Through their National Chapters they bring together people of integrity in civil society, business and government to work as coalitions for systemic reforms. As they outline in their Mission Statement they do not identify names or attack individuals, but focus on building systems that combat corruption. They are playing an important role in raising public awareness and their Corruption Perceptions Index has triggered meaningful reform in many countries.

Transparency International classifies countries according to their level of corruption, giving Denmark the grade 10, or first place, for being practically without corruption, and to the other Scandinavian countries, New Zealand (3rd 9.4), Canada (5th 9.2), Singapore, the Netherlands (8th 9.0) and Switzerland (10th 8.9), the ten least corrupted countries of the world, with notes of 10 to 8.9. Australia is 12th – 8.7, Great Britain is 13th - 8.6 and Germany is 14th - 8. The three countries analyzed in this book are more or less at the same level of corruption, with the United States in 18th place (7.5), Israel in 20th place with 6.8, and France in 22nd place with 6.6. In three years Israel has deteriorated from 14th place to 19th place and in 1999 to 20th place. Japan receives the grade of 6 in 25th place, Belgium 29th – 5.3, South Africa 34th – 5.0, Italy – 4.7 in 38th place, and Russia – 2.4 in 83rd place. Cameroon is the most corrupted country with 1.5 in 99th place.

"In spite of the more active role of the press; the more perseverant acts of the judges; the more persisting pressure, opinion and intentions of the government, always proclaimed convincingly, France has not really succeeded to cure itself of this illness that eats away our societies. In the classification established by TI, our country appears every time in a bad posture. The French companies continue to consider bribes as necessary to win a contract: only Japanese and Italian companies in the western world are more lenient. Everywhere else, in Sweden as in Great Britain, in Germany as in Spain, they are more conscious and disposed to tackle the matter. Corrupting, our country is perceived as corrupted. In order to get contracts in our country, it is necessary sometimes to resort to unthinkable practice. If Denmark, Finland, New Zealand and Canada appear as the countries more respectful to a certain business ethics, France and Spain are ranked only in 22nd place. Among the 15 of the European Community, there are only three in which apparently the sickness is even deeper – Belgium, Greece and Italy.

Beyond this statement, the action taken by the Paris government in its fight against corruption appears to be very shy. The anticorruption convention of the OCDE (l'Organisation de cooperation et de developpement economique, the rich countries club) in 1997 can be criticized: it does not tackle in effect all the links of the chain. It does not refer, in particular, to the issue of the fiscal heavens – a theme over which the French Minister of the economy, Dominique Strauss-Kahn, insists rightfully, every time he participates in an international convention. This OECD treaty is nevertheless a progress in the indispensable international cooperation. But even on such a modest agreement, France is behind schedule. In contrary to most of the other exporting countries, it has not yet ratified the agreement; it has not yet introduced the principal elements of these commitments in its own legislation. Before lecturing others, one has to give the good example himself." (Le Monde, 28.10.99)

ADAM

Corruption and lack of ethics are closely related, and the most striking lack of ethics prevails probably in the relations between companies and minority shareholders. "There is often a confusion of terminology between 'minority shareholders' and 'individual shareholders'. Evidently, the individual shareholders are minority shareholders, at least when they are taken individually, but the minority shareholders are principally those who do not control the companies. They are those who invest capital – investors – who incur the risks to enter as shareholders in companies that go public, with a goal to share the obtained profits. In France, the individual shareholders (more than five million) hold about one third of the capitalization in the stock exchange.

They weigh therefore as heavily as the foreign investors (essentially Anglo-Saxons) who hold another third, and three times more than the French O.P.C.V.M. that hold about 10 percent of the shares capitalization. In total, it is therefore at least three quarters of the stock exchange capitalization that are held by minority shareholders. A phenomenon, linked certainly to the privatizations that have opened the capital of the French companies to millions of small shareholders as well as to foreign capital. But in proportion to the GNP, the stock exchange capitalization is still much inferior to its level in the Anglo-Saxon countries. In comparison to their foreign competition, many French companies are fragile and their growth is slowed down by the lack of equity, especially after the stop of inflation which does not allow them to borrow massively to get financing. They have therefore to raise money from the shareholders." (Neuville, Protection judiciaire des actionnaires minoritaires, legal protection of minority shareholders, p. 1-2)

The shareholders have to be given reliable information in order to reach decisions, which is not always the case, as we can learn from a large number of cases that were published in the press in recent years and that were treated by ADAM and its president Colette Neuville. ADAM - l'Association pour la Defense des Actionnaires Minoritaires, the Association for the Safeguard of the Minority Shareholders - was created in 1991 by its president, and has succeeded in obtaining an outstanding place in the media in its campaign against wrongdoing to minority shareholders. The goal of ADAM is to help improve the functioning of shares companies, which is inseparable from the protection and development of savings invested in companies. ADAM is open to individual shareholders and to organisms with collective investment, French or foreign, that wish to join forces in order to safeguard their rights. ADAM tries to promote the surveillance and control of the shareholders over the executives of the companies in order to insure that they will have as objective only the maintaining of the common interest of the shareholders, according to the corporate law (among others article 1833 of the Code Civil) and corporate governance.

ADAM has intervened, inter alia, in the following cases:

End of 1991/92, in order to defend the minority shareholders of Printemps during the partial O.P.A. of Pinault over Printemps.
During 1992, in order to obtain the increase in the exchange parity in the O.P.E. of Suez over C.F.I.
In 1993, in order to obtain the increase in the price of the public offering of Corela at the change of majority shareholder of the parent company.
In 1993/94, in order to preserve the rights of the holders of convertible debentures at the restructuring of the debt of Eurodisney.
In spring 1994, in order to defend the minority shareholders of La Redoute at the merger by Pinault-Printems.
In 1995, in the Sogenal affair where ADAM has tried to establish the rights of the expropriated shareholders on the net equity.
In spring-summer 1996, in order to obtain from the state an O.P.A. in favor of the shareholders of Credit Foncier de France.
In 1997, ADAM has defended the shareholders of Havas who wish to obtain an equitable exchange parity during the operations of restructuration of the group of Generale des Eaux.

On the other hand, many companies and university professors maintain that minority shareholders harass the companies in order to extort benefits that are not due to them, claiming that some minority shareholders are 'speculators who are eager to have prompt benefits and have no respect and loyalty toward the companies where they invest'. Therefore, according to them, it is their duty to prevent their schemes by forcing them through the courts to pay damages. If the issue of the minority shareholders is tackled under a strictly defensive angle, we can find many cases in which minority shareholders resort to harassment maneuvers that not only destabilize the management in charge but can also in due term threaten the social interest. As the right to criticize that is recognized for the minority shareholders has only a goal to serve strictly their individual interests, the protest becomes pure harassment reprehensible as other sorts of harassment, such as contractual harassment. Those strategies of harassment have sometimes received some encouragement, notably through the decisions of the Court, especially in cases of class actions.

It is against those alleged harassments that those companies try to protect themselves and the judges justify the companies if they are convinced that the matter is in fact an abuse by the minority shareholders. The minority rights are not evident, even from a legal point of view, and the minority shareholders, who do not want to risk being sued for harassment, have, in the end, only ethics to safeguard their interests. It is very difficult for minority shareholders to prove legally that their rights were wronged. On the other hand it is much easier for companies to prove the opposite case of harassment by minority shareholders, those despicable 'speculators'.

In France, there are legal procedures that allow the sanctioning of harassment of shareholders. The judges have the power to sanction by fines the procedural moves that are too abusive. Various laws allow the sanctioning of the abuse of procedures: article 91 of the Code de procedure penale permits the sanctioning by a civil fine of unfounded complaints; it is similar in the civil law, where a maximum fine of 10,000 F could sanction an abusive procedure and article 700 of the new Code de procedure civile that allows in a certain way to 'make the bad plaintiff pay'. The abuse of the minority shareholders is sanctioned by the attribution of damages, a normal compensation method for the suffered harm. The condemnation to pay damages match notably a certain number of decisions to reject requests for information on behalf of the article 226 of the law of 1996.

OTHER ACTIVIST GROUPS

Arese, Analyses et Recherches Sociales et Environnementales sur les Entreprises, the French organization that researches the ethical and environmental conduct of companies, headed by Genevieve Ferone, has ranked the top European companies screened by different criteriae. Thus, in Human Resources, Employees Relations, the best companies are: Ahold, Electrabel, Endesa, Hypovereinsbank, Mannesman, Metro, Philips Electr., RWE, Unilever and Veba. In Preservation of the Environment the best companies are: Ahold, Daimlerchrysler, Mannesman, Metro. Nokia, Philips

Electr., Royal Dutch Petroleum, RWE, Siemens, Unilever. In Customers and Suppliers Relations the best companies are: BCO Bilbao Vizcaya, Dresdner Bank, Electrabel, Endesa, Hypovereinsbank, Nokia, Philips Electr., Repsol, Siemens, Unilever. In Shareholders Relations the best companies are: Ahold, Daimlerchrysler, Dresdner Bank, Electrabel, Metro. Nokia, Philips Electr., Royal Dutch Petroleum, RWE, Unilever, Veba. In Relations to the Community and the State the best companies are: Ahold, BCO Bilbao Vizcaya, Electrabel, Endesa, Hypovereinsbank, Philips Electr., Repsol, Royal Dutch Petroleum, RWE, Unilever.

The following is a sample of social investing and consumer activist groups and organizations: 20/20 Vision for protecting the environment, Action against Hunger, The Action Coalition preserving human rights, the American Animal Care Foundation, Center for Biological Monitoring, Center for Defense Information monitoring and criticizing the military, Center for Economic Conversion, Computer Professionals for Social Responsibility, Co-op America provides practical tools for businesses to address social and environmental problems, Council on Economic Priorities, Cruelty Free Investment News, Earth Challenge, Earth Wins, Environmental Defense Fund, The Equality Project, Fair Trade Federation, Friends of the Earth, Grassroots International working for social change, Habitat for Humanity International, Hunger Web, Inner City Press on community reinvestment, Institute for Global Communications, International Co-operative Alliance, International Federation for Alternative Trade, Macrocosm USA for urgent social and environmental problems, New Uses Council for new consumer uses of renewable agricultural products, New World Village for the politically progressive Internet community, the Nonviolence Web, Nuclear Information and Resource Service, Pax World Service, Physicians for Social Responsibility, The Progress Report, Public Interest research Groups, Rainforest Alliance, Social Justice Connections, Union of Concerned Scientists, Zero Waste America.

An alternative to safeguard the interests of minority shareholders could be the progressive taxation of profits in order to decrease the greediness of the shareholders and share in a more equitable manner the profits among the various shareholders. We could also tax mainly the short-term profits, made by buying and selling shares within less than a year. In this manner we could promote the ethics of the transactions by diminishing the drive to make frauds, excessive profits, and short-term profits. We need to change the image of the casino, which has received today the stock exchange, to an image much more serious and conscientious, which will attract ethical shareholders and not speculators. We should alienate from the stock exchange people eager to make prompt and excessive profits to the detriment of the other shareholders, who benefit from insider information and engage in fraudulent activities, being encouraged by the extraordinary profits.

Ivan Boesky has declared, before his fall, to the graduates of UCLA that you can be greedy and feel good about it. Perhaps if the temptation would not be so great, he and others would have abstained from their fraudulent conduct and he would have continued to respect the law. We remember what Aristotle has said on moderation and if we could make a wish to the minority shareholders it would be that they would have moderate and not excessive profits, as this is the best safeguard for them and for the majority shareholders to make ethics prevail and keep their integrity, even when they will participate in the control of the companies!

Karl Marx did not believe that the proletariat existed as a class conscious of its rights when he wrote 'Das Kapital'. The minority shareholders, nowadays like the proletariat in the 19th century, are not associated and conscious of their power. Marx has noticed the excessive abuse of power of the capitalists of his time who managed the economy not with the invisible hand of Adam Smith but with an iron fist, which oppressed the masses. It is Dickens, Zola, Hugo and others who have described the sufferance of the masses, but unfortunately modern literature does not pay attention to the wrongdoing to minority shareholders. Marx and Zola have condemned the indifference and injustice of the mighty toward the poor, the weak, those who were not organized.

"Taking the labor theory of value to its logical conclusion, Marx argued that those who did the work produced the value and, consequently, deserved the products of their labors for themselves. In other words, his emphasis on the actual activity of production instead of the commercial value of the end products led him to a conclusion that would have not been tolerable to Adam Smith – that the work itself was everything and the operations of the market were only a systematized form of theft. Marx, in other words, is very much in the line of ancient and religious thinkers who rejected the activity of business as parasitic on the honest labor of the working man... That concept is exploitation, and it is the sense of being exploited that did, in fact, create the class consciousness Marx urged (for example in the American labor union movement) and that continues to appeal so powerfully to so many people in Third World countries, especially former colonies of the great industrial empires." (Solomon, Above the Bottom Line, p. 267)

Nobody advocates to end up with the conclusions of Marxism in order to safeguard the interests of the minority shareholders, although the basic situation is the same – they are the majority of people contributing the most to the economy but sharing only a fraction of their contribution without being represented adequately. The solution should be a cooperation between the majority and minority shareholders and the management of the companies. But in order to reach this stage, it is needed that the minority shareholders should sense that they are despoiled in many cases, they should organize in order to safeguard their interests, they should be assisted by the activist associations. We could do it by way of evolution or by revolution. The powerful should reach the conclusion that it is in their best interest not to abuse their excessive rights, exactly like in Great Britain, which has managed to move from absolutism to democracy without revolution.

The revolution for the minority shareholders would be to cease investing in the stock exchange, after having lost their trust in the system. The minority shareholders have the alternative to invest their savings in the banks instead of purchasing shares of companies. They could earn much less, eventually, but they would not incur the risk of being despoiled in fixed games, by fraudulent use of insider information, and by greedy businessmen eager to get even richer at all cost. But if the minority shareholders cease to invest, the stock exchange will suffer from its worse collapse ever, which could end up in a world recession. If we do not want to encounter such catastrophes, we should allow the minority shareholders to exert their rights, share equitably in the companies' wealth, be represented adequately in their organizations, participate in their control, and restitute the notion of fair play in the stock exchange.

8
CASE STUDY OF THE FRENCH COMPANY LOSKRON

"Now a traveler came to the rich man,
But the rich man refrained from taking one of his own sheep or cattle
To prepare a meal for the traveler who had come to him.
Instead, he took the ewe lamb that belonged to the poor man
And prepared it for the one who had come to him."
David burned with anger against the man and said to Nathan,
"As surely as the Lord lives, the man who did this deserves to die!
He must pay for the lamb four times over,
Because he did such a thing and had no pity."
Then Nathan said to David: "You are the man!"
(The Bible, 2 Samuel, 12:4-7)

In the cases of this book the names of the companies, individuals, newspapers, etc. are fictitious, in order not to reveal the identity of the persons and the companies. Nevertheless, all the cases are backed by adequate documentation. As this book is being published and its author presents his beliefs on the conduct of the companies and their executives and owners, it was found preferable to generalize the cases, as ultimately, the book is not intended to obtain justice in one or all of the specific cases, but to examine in concrete cases the ethical conduct toward minority shareholders.

The Loskron case is presented in different angles, the first one describes the company as the victim and the minority shareholders as greedy speculators. "Spectacular as well is the decision of the tribunal of commerce of Lirotts from July 7th 1995. All the minority shareholders are condemned to damages of 400,000 F. We have to add to that an article 700 of 50,000 F. The complaint of the minority shareholders was aimed to compensate the harm resulting from an operation of the type of 'coup d'accordeon'. The motivation of the judgment is worth being examined as it is very symptomatic from the point of view of the subject that preoccupies us: 'Since it has to be noted that the lawsuit of the minority shareholders was filed through the investigation and the active moves of ADAM, which has become a shareholder of the SA only lately and that took a predominant role in the contentious strategy; that it is worthwhile also to mention the action of the company Creon... which, conscientiously, buys 5,231 shares for 1F each, knowing that it will lose them irrevocably three days later; that the court will allocate the largest part of the responsibility of the abuse of rights of the guilty minority shareholders of the SA to those two plaintiffs that have acted only for sheer speculation, at the expense of a company that its main concern was to avoid the disappearance of the company and its related jobs". (Cossev, Le harcelement des majoritaires, harassment of majority shareholders, p. 117, Bejov, February 1996)

It is always easy to draw retrospective conclusion, a few years later, when the intentions and the results of all parties concerned are clear and evident. But what do we have at the date of the judgment against ADAM and the minority shareholders of the company Loskron. One can think that the minority shareholders and ADAM are greedy speculators that have to be punished severely for their harassment of the company Loskron that has acted by sheer concern to avoid the disappearance of the company and its jobs. The patriotic company, an exemplar citizen, is harassed by those speculators that have only one interest – to maximize their profits at the detriment of the employees of the company. 'Abuse of rights', 'guilty of', 'wishing to speculate', 'at the detriment of a company that has only one concern – to keep the jobs of its employees'. ADAM buys therefore shares of the company a few days before they cease to exist in order to speculate and 'insure a predominant rule in the contentious strategy'. What is therefore this perfidy, this ruse, this scheme hatched up by greedy and well-known speculators? To describe this case we have to start from the beginning and explain how we arrived to this stage and describe all the peripeteia of this 'intrigue'.

The company Loskron is a French company, which manufactures and sells fashion accessories. The founder of the company was born in the eighteenth century and in 1810 his company already exported products to Italy. In 1883 – a modern factory was built. In 1963 – it adopts its present name and increases its workforce to 120 persons. In its market segment, the company is the undisputed leader and it manufactures and sells products of the most well-known brand names. In 1998 it employs more than 800 employees and has a sales turnover of 507M francs, with a profitability of about 10M francs. Its principal markets are Europe – 70 percent and the United States – 21 percent.

But the situation was not so bright in 1994, when the company incurred losses that jeopardized its existence. The company Loskron was quoted at the stock exchange and at the eve of the annulment of its capital, on 8.8.94, the majority shareholders held 88 percent of the shares and the minority shareholders – 12 percent (while on 22.7.94 the shares distribution was 75 percent to 25 percent).

A few months earlier, on May 17, 1994, the French court, Cour de cassation, authorized in another case the annulment of the capital of a company. This was done in a case where the net equity of the company ... reached less than half of its invested capital. The special general assembly of the shareholders had, according to the corporate law, to decide on the continuation of the company's activities. In such an event, the shareholders have the option to decide on the advanced dissolution of the company, or the reconstitution of the net equity in order to enable the continuation of the activities. The goal of this reduction of capital is to absorb the debt of the company and following this 'coup d'accordeon' – 'accordion action', all the losses are written-off and the company can clean its balance sheet and continue its activities after having reassured its creditors. The shareholders have lost their investment in the first reduction of the capital, thus playing fully the role that is assigned to them by law and by the company contract, by contributing to the losses of the company.

When the equity is exhausted, the activity stops, unless one or many shareholders, former or new ones, agree to invest more money into the company. It has happened to ..., that went bankrupt. The small shareholders who contributed to the creation of the newspaper were informed that they have lost their investment and that the management of the newspaper was looking for new partners. For ..., the former shareholders were given the opportunity to participate in the new investment, which showed their affectio societatis. In fact, most of the recent cases give a priority to the old shareholders to subscribe to the new offering. Only one exception, the case of Loskron, in which the new investment was proposed only to an external investor. The judgment of the Cour de cassation of 1994 permits the annulment of all former shares when reducing the capital to zero, thus making the former shareholders disappear. From this moment on nobody has any more rights. Even if the COB (the French SEC) encourages the companies to give to their former shareholders a priority right to reinvest in their company, it is only a favor but not a right, especially when this option could dissuade a potential investor who does not want any partners in the control of the company.

It is precisely in this context that this case will analyze the rights of the minority shareholders and examine if they have been wronged. On August 8, 1994, the shareholders of Loskron were excluded from their company by a decision of a shareholders' meeting. Considering as unlawful the procedure that excluded the minority shareholders from their company, ADAM decided to contest the right to do this action in order to prevent the establishment of a precedent. At this date the shares of Loskron have been quoted for many years and have obtained very high prices that reached even 900 francs per share.

"On ... 1986, Loskron introduced in the 'second marche' of ..., the second stock exchange market in France, 10 percent of its capital (53,300 shares) at the price of 210 francs per share. But trade was impossible. 7.6 millions shares are requested. It is only on … that the share can be quoted at the price of 335 francs. Only a year after, in December 1987, it reaches 900 francs. Those that had the good idea to sell have done a good deal, although many analysts still recommended to buy. It is true that the prospects for 1988-1989 seemed very favorable, with profitability increase of 15 percent. But it was necessary to be brought down to earth. Since 1989, results are deteriorating. It brings the company in 1993 almost to bankruptcy with 120 millions francs in accumulated losses and 250 millions francs of debts." (Des yeux pour pleurer, Eyes to Weep, Vesno, 20.8.94)

In order to give an idea of the losses of the shareholders of Loskron, one can bring the example of M. P who bought on 25.3.91 18 Loskron shares at the price of 348 F per share, on 2.1.92 – 28 Loskron shares at the price of 222.3 per share, on 6.11.92 – 50 Loskron shares at the price of 100 F per share. The Loskron share was quoted at the stock exchange on 4.1.93 at the price of 65 F, and a few days later the price was more than doubled. A few months later, the price falls back to the level of early January 1993 and stays at this level until its suspension, seven months before annulling the capital.

Quotes of the Loskron shares are resumed on July 26, 1994, and the British company Kepler (later on renamed Epsoks), that buys Loskron, purchases almost all the shares at a unit price of one franc, with 13,000 transactions. Shareholders who have bought their shares at the stock exchange for a price ranging from 65 to 348 or even 900 francs, have practically lost all their investment and cannot participate in the new investment in the company Loskron. If they do not sell their shares to the buyer for 1F, their shares will be annulled and they will not be able to benefit from the fiscal benefits of the loss of their investment by reporting a deductible fiscal loss.

Apparently the family Loskron loses all its capital without getting anything and the company is bought by the American Frolvos. "At the age of 29, the New Yorker Frolvos has bought Loskron, the French no. 1 of … This young man in a hurry, who is a relative of …, has built in two years a group of 1.5 billion francs of revenues, quoted at Wall Street. He has seen a lot of documents on companies in trouble! But 'this one was much more rotten', he assures. It is true that, with 221 millions in losses and a sales turnover down by 20 percent last year, Loskron is in a bad shape… Loskron has launched in the beginning of the '90s an aggressive expansion plan: strengthening of the control, opening of sales offices abroad, buying of competitors. Those expenses have brought down the company's profitability in 1992, without increasing its sales. Nevertheless, the patriarch of …, who has transformed the factory of his father to a small multinational with 1,500 employees, persisted. It was out of the question to give the management to an external manager and to open the company to investors who might strengthen the financial situation. The company will remain family owned. By the end of 1993, the situation was unbearable. Loskron survives only with bank indebtedness, and the shares' quotation at the stock exchange is suspended on ... at the level of 65 francs. The interministry committee of industrial restructuring (Ciri) interferes in order to save the first employer of ... from bankruptcy.

Mission accomplished. On August 8, 1994, Loskron is bought without going bankrupt. L'honneur est sauf, appearances are kept up: bankruptcy is avoided, and the nephew of ..., Jean Loskron, will remain as CEO. But, otherwise, what a mess: the family disappears from the ownership without receiving a penny. The minority shareholders see the value of their shares reduced to one franc. Only the public authorities admit discretely their satisfaction: 'The important thing is to succeed in saving employment; it will not be shouted over the roofs that the banks were ripped off.' Those, mainly Celine, have indeed lost two thirds of their 215 millions francs of debt to Loskron. Delighted from the opportunity and full of admiration to the family Loskron, 'that did not recoup anything and has lost everything', the American savior has probably achieved the best deal of his career. As the company, in spite of its disclosed deficit of 221 million francs, is much more seductive than it looks. The operational loss was limited to 12.2 millions francs last year, twice less than in 1992. The license contracts have been renewed, and the subsidiary ..., one of the best of the group, has doubled its results in 1993. The clean-up has been done everywhere, and the Loskron have paid for it in full. Besides the lay-off plans, the hole made by the unsuccessful acquisitions of the beginning of the '90s has been filled up. On the results of 1993, for example, 80 million francs are allocated to the closing down of two companies purchased in 1991. In the 1993 results of Loskron, we find 21 million francs of provisions to 'conform to the Anglo-Saxon accounting standards' or 59 million francs for future restructuration. Provisions, that have brought up 'reserves' from the auditors, as they should be in principle borne by the buyer and not by the seller.

For the Loskron, the most important thing was to abandon the ship in order. Well, almost. As, unlike the public statements, no guarantee for employment was insured to the 875 employees of the company. The psychosis of the lay-offs is still there. During the shareholders assembly of August 8, a Parisian lawyer representing the interests of the buyer has stated that there were no legal engagements on this subject. Joined in New York, Frolvos is formal: 'It will be a suicide to guarantee employment if sales decrease.' The former owners were therefore mislead by eliminating any solution of a takeover by a rival fortune... on the pretext that they did not have sufficient financial background? Frolvos, although being a New Yorker capitalist, has bought Loskron without a penny. In order to pay 75 million francs to the creditors and raise the funds lacking to this new 'European bridge', he intends to raise 150 million at the London Stock Exchange. With the assistance of the Celine Securities... (A vouloir rester familial, Loskron est devenu americain, by wanting to remain family-owned, Loskron has become American, p. 61, Noslan – No 958 – 12/08/94)

We can therefore detect a logic in the intrigue, that will be repeated in the other cases as well. First of all – there is a company that is in a very bad financial shape, real or apparent. The buyer and seller are suspected of coordinating their tactics and give the impression that the company is in such bad condition that nobody else is interested in it. The buyer says: 'companies in trouble, I have seen!' But 'this one is the most rotten'. Then, the banks act in some cases in cooperation with the parties, as they are interested in recouping a large part of their loans that otherwise would be completely lost. The bank Celine forgives two thirds of their 215 million loans to Loskron. Why is it that the banks help the parties, is it truly to help a client in need, to recoup the remainders, to prevent the lay-off of the employees of the company?

Or maybe they have a 'hidden agenda'? In the cases of the Israeli companies Furolias and Erinsar, the banks are directly or indirectly shareholders of the companies. In the case of Loskron, Frolvos 'although being a New York capitalist has bought Loskron without a penny'. A similar case to Furolias, where Erinsar buys Furolias with a shares offering based on the acquisition. The buyer of Loskron pays only 75 million francs to the creditors, but he raises for that purpose 150 million from the London Stock Exchange, with the assistance of the same bank Celine. On the 23.12.98 it is still the same bank Celine that gives its opinion to the independent directors of Epsoks to accept the takeover bid of Jean Loskron to repurchase the Loskron company that was sold in 1994 to Epsoks. The independent directors consider the OPA (public offer to purchase the shares or takeover bid) as 'fair and reasonable'. This is another common characteristic to Furolias, Erinsar and Soktow, as well, where the independent directors decide almost always according to what the controlling shareholders wish, being nominated by them. This might bring us to the conclusion that as long as the independent directors will not be nominated by an independent body the rights of the minority shareholders will continue to be wronged and the independent directors will not rescue them.

In the analyzed cases, the buyers make an excellent deal, as they buy a company with a great potential for a ridiculous price. They buy the company on the verge of return to profitability, when the crisis is already over. The minority shareholders are told that the company is not worth anything and they are not given anything, but Frolvos receives at a bargain price the Loskron company, with a formal deficit of 221 million, but with a reduced operational loss of 12.2 million francs last year, half of that in 1992. The license agreements were all renewed and the hole made during the unfortunate acquisitions of the beginning of the '90s is filled up. We can notice that the losses are calculated in a very conservative way just when there is a need to sell the company in order to give an impression of a catastrophe. In the 1993 results, 80 million francs are allocated to the closing down of two companies acquired in 1991. We find also 21 million francs as provisions to 'conform with the Anglo-Saxon accounting standards', and the extreme is the 59 million francs for future restructuring. This tendency is common to all the other cases analyzed in this book. We take maximum 'provisions', we give the impression that the company will go bankrupt, the majority shareholders allegedly cooperate with the buyer and the banks, and the only ones who are wronged, as they are the weakest, are the minority shareholders. It is true that the auditors have 'reserves' on those proceedings but they do not resign and finally they collaborate with the companies.

It is necessary to validate those arguments, as, apparently in the case of Loskron, the majority shareholders lose all their investment, exactly like the minority shareholders and they sell their company uniquely in order to safeguard employment in a region where they have managed their company for more than 200 years. But, miracle, a few years later, they buy back the company! Jean Loskron, keeps his position as CEO of the company, although it was sold, a very rare case for a CEO who was with his family the majority shareholder and who sells all his shares. And, another miracle, the Anglo-American buyer invests in the company a much larger amount than what he receives five years later for selling his shares, although the financial situation of the company has much improved in the meanwhile. This is totally incomprehensible conduct for a businessman as shrewd as Frolvos. It is beyond the scope of this book to analyze the motives of the buyers and sellers, we can only judge by the results, which are very favorable for the Loskron family, who has succeeded in getting back its company, while the minority shareholders have lost all their investment.

The sequence of events is therefore as follows: "After having fought for more than a year to save the family-owned company, Jean Loskron can now devote his time to turnaround the company that bears his name, and that this summer the Anglo-American group Bacon has saved from bankruptcy. While the family has no more shares in the Loskron company, henceforth owned in 100 percent by Epsoks, Jean Loskron is still the CEO together with Frolvos, CEO of the American holding Bacon and Chairman of the board of directors of Loskron. The changing of ownership has enabled the improvement of the results of Loskron, which had an accumulated loss of 220 million in 1993 with 75 million net deficit. After the accordion action approved on August 8, the company Epsoks (the new name of Kepler) has invested 80 million francs in equity in Loskron and purchased its bank debts, all in all 175 million francs after discount. This sum has to be transformed in quasi-equity in the following months. A working capital investment of 40 million francs will be made within two weeks. Bacon finances this recapitalization of Loskron, by issuing its British subsidiary, Epsoks, that has raised 150 million francs, specifies

Jean Loskron. 'Loskron will be breakeven this year and will reach a 5 percent profitability in 1996' he says." (Loskron repart sur des bases assainies, Loskron starts again on sounder basis, Ertak, 15.9.94)

The buyer pays therefore 175 million francs, that he gets almost completely from a shares offering at the London Stock Exchange for part of the shares of its British company Epsoks. The money is invested in the company Loskron and helps to pay back part of its creditors. And, as if by magic, a month after the minority shareholders of Loskron were told that their company was almost bankrupt, Jean Loskron the CEO of the company, that maybe wants to facilitate the issue of the shares of Epsoks, declares that the Loskron company will be breakeven in 1994 and will reach a 5 percent net profitability in 1996. But nobody distributes shares of the British company to the minority shareholders, and lets them benefit from the turnaround of the company, as they have already lost all their investment. The minority shareholders, we recall, have bought their shares for 65, 100, or 222 francs, while the losses of the company were very heavy and are obliged to sell their shares for one franc or lose all their investment, while the company succeeds to breakeven in the same year. This logic is very important to be understood, as it is repeated exactly in the same time in Israel in the case of the company Furolias. There are therefore two different scales and measures, one for the minority shareholders and one for the others.

But what happens ultimately with the majority shareholders, the Loskron family? They manage to buy back the British company Epsoks in 1999 for 130 million francs, much less than what the British, or more specifically the 'minority shareholders', who have bought 74 percent of the shares of Epsoks at the London Stock Exchange, have paid to purchase the French company Loskron. The control of Epsoks has remained with Bacon, that held following the shares issue in London – 26 percent of the shares of Epsoks. And what happened to the price of the shares of Epsoks? The valuation of the company was in 1993 – 10.2 million sterling pounds, it increases by more than 150 percent in 1994, after the purchase of Loskron to 26M, goes down to 24.5M in 1995, to 11.5M in 1996, 11.1M in 1997. Finally the shares price varies between 10 to 20 p. in 1997 and 1998, and the last quoted price of Epsoks, that has 88,610,000 shares, at the stock exchange on the 22.1.99 is 14.5 p., just before the purchase of Epsoks by Jean Loskron. The valuation at this price amounts to 12.8 million pounds. In five years the valuation of the British company did not increase actually in real terms, although it has purchased the French company Loskron, and made a remarkable turnaround.

The British company that sold four million pounds in 1993 with losses of 0.25 million has distributed a dividend of 1.2 p. per share in 1993, 1.5 p. in 1994 with sales of 21 million and losses of more than three million, 1.65 p. in 1995 with sales of 61 million and profits of three million, 0.57 p. in 1996 with sales of 59 million and profits of 0.5 million, and no dividends in 1997, although the sales in this year amounted to 57 million pounds and the profits were above one million pounds. The British shareholders could not understand how a company with profits of 3.2 million pounds in 1995 has only a profit of 0.5 to one million in the following years with almost the same sales turnover, unless the profits in 1995 were mainly generated from the excessive reserves of Loskron in 1994 that have caused the reduction of the capital to zero and the annulment of the shares of the minority shareholders. The only thing that they know is that the valuation of the British company was reduced by 70 percent in two years.

This is why the British shareholders have accepted the offer of Jean Loskron, although their company sold 14 times more than in 1993 before the purchase of Loskron, was profitable, but had the same valuation as in 1993 with no distribution of dividends, while it distributed dividends of 1.2 p. per share in 1993 and 1.5 p. in 1994 when it incurred heavy losses. But the British shareholders, who do not have an organization similar to ADAM, have not retorted and have suffered their fate without complaining, although they have among their minority shareholders very large funds. Part of these shareholders have bought their shares in 1995 at 30 p. and in 1996 at 45 p., and were forced to sell their shares at 15 p. in 1999, incurring a loss of 50 percent to 70 percent.

Jean Loskron buys back his family company and its British parent company for the sum of 130 million francs, while the French company has a sales turnover of 507 million francs and a profit of 10 million francs. We remember that in 1994 Epsoks has purchased with the money of the new British shareholders the company Loskron with a deficit of more than two hundred million for the amount of 175 million francs. But even the purchasing's lower amount of 130 million francs is financed by a loan of 50 million francs, the remainder being financed half by shares and half by convertible debentures to shares. How did Jean Loskron succeed in financing the equity needed for the purchase of Loskron, by notes that he may have received from Epsoks or by another method?

Ultimately, Jean Loskron has made an excellent deal and has bought back his company, that he has continued to manage in the last five years, at a very attractive price... "La boucle est bouclee, the buckle is buckled... The group Loskron, the first French manufacturer of ..., almost bankrupt in 1994, was bought at this time by an American financier Frolvos, with the assistance of Anglo-Saxon funds. The PME was then located in the British company Epsoks, quoted at the London Stock Exchange. At this time, Jean Loskron from the family who founded the company, has remained the CEO, with Descartes, the CFO. Five years later, after a sharp reduction of workforce and industrial sites, the company is again a French company. Jean Loskron and

Descartes launched by the end of last year, via the intermediary of an ad hoc company, Kofna, an OPA called friendly over Epsoks at the London Stock Exchange. This offer was successful, as in January 22, Kofna held 93 percent of the capital. The whole amount of the OPA, at 15 pence per share is in the order of magnitude of 130 million francs. The takeover is financed by a loan of 50 million francs, the remainder being financed half through shares (allocated to the two French managers of Loskron) and half through convertible debentures into shares, subscribed by Uriel, the company of Clavaux. Loskron, which employs more than 800 employees, has made in 1998 a sales turnover of 507 million francs with a profit of 10 million francs. (Loskron passe de nouveau sous pavillon francais, Loskron is once again a French company, Ertak, 19.2.99) The only partners that were wronged are therefore the French and British minority shareholders!

This case emphasizes one of the inflexible laws of the business world of the end of the century: Finance the investments through minority shareholders while maintaining control of the companies, with a core that varies in general between 20 percent to 40 percent. In this manner we have noticed how the American company Bacon, controlled by Frolvos, buys the small British group of Kepler, called now Epsoks. Bacon sells 74 percent of the shares of Epsoks at the London Stock Exchange in order to finance the purchase of the French company Loskron. Bacon keeps only a core of 26 percent that enables it to manage and control the company, as all the other shareholders of Epsoks have only each less than 10 percent of the equity.

We have learned what was the fate of the British shareholders. But it is impossible to learn from the newspapers, the publications of the company and the Internet, what was the personal profit of Frolvos from the acquisition of Loskron at a higher price than the selling price, or what is the breakdown of the control and the shares among the Loskron family. The company of Frolvos is one of the American leaders in the product line of the Loskron company, which is itself the French leader. Bacon could have received indirect benefits issuing from the cooperation, which exceed the profitability of the shares' transaction. On the other hand, we know that Jean Loskron has taken the position of CEO in Loskron only in June 1993 and he managed to structure the company in such a way that he rescued it from bankruptcy. It was he who remained the CEO of the company, during the whole period of the acquisition by Bacon and its subsidiary Epsoks, and who bought back the company with his CFO in 1999. The other members of the family had or did not have shares of the company in 1994? What have they lost in the transaction? It is impossible to know, as those family transactions are not published. The banks and creditors of Loskron could have received notes of the British company in order to compensate them on part of their losses in the forgiving of Loskron's loans. Therefore, we are in doubt about part of the partners, such as Frolvos, the Loskron family or the banks and creditors.

Only, as far as the minority shareholders are concerned, we have the certainty that they were wronged. It is therefore, in order to safeguard their interests, that they have approached ADAM.

Some of the minority shareholders contacted ADAM in 1994 and asked it to help them recoup part of the losses that they have incurred. On 15.11.94, they serve to the Loskron company a summons to appear before the tribunal of commerce of Lirotts. They do not request the annulment of the decisions taken on 8.8.94, as such a move could jeopardize the future development of Loskron and the loss of more than 800 jobs in the company. The summons describe the facts as follows:

"2. On July 4, 1994, the majority shareholders of Loskron sign an agreement with the company Kepler by which they commit to vote in an extraordinary shareholders meeting first of all on a reduction of the capital by annulment of all the existing shares and then an increase of capital reserved to Kepler. The minority shareholders of Loskron learn about the existence of this agreement by the press of July 1994. On August 8, 1994, the extraordinary shareholders meeting decide:
Second resolution '... under the suspensive condition of the fulfillment, prior to August 20, 1994, of the increase of capital treated in the fourth resolution, to reduce the equity of the company by 17,563,920 FF, for bringing it from 17,563,920 FF to 0 FF in order to eliminate the negative equity of 141,446,311 FF. Third resolution '.... To fulfil the reduction of capital by way of annulment of the existing shares'. Fourth resolution '... to increase the capital by 80,000,000 FF, by the issuing of 800,000 new shares of 100 FF each, issued at par value, without an issuance fee, to be delivered in total in cash or to compensate liquid and requested debts to the company, at the issue'. Those resolutions, justified by the losses incurred by Loskron SA, do not exclude as such the shareholders: in fact, those retain their standing as shareholders and as such are allowed to participate in the increase of capital treated in the fourth resolution.

3. In the fifth resolution, the majority shareholders of Loskron SA decide: 'The extraordinary shareholders meeting, in view of the terms of the report of the board of directors and the special reports of the auditors, decide to annul the preferred application for shares right in favor of the company Kepler (whose name will become Epsoks) that will have the exclusive right to subscribe for the 800,000 new shares which will be issued for the increase of capital treated in the previous resolution. Consequently, the extraordinary shareholders meeting states that the adoption of the present resolution, regarding the abolition of the preferred right to subscribe, results, in view of the reduction of existing capital to zero in the second resolution, in the automatic exclusion of all the existing shareholders of the company in the context of the operations of reduction and increase of capital treated in the

previous resolutions'... The exclusion results in the <u>conjunction</u> of the reduction of capital to zero and the reservation to Kepler of the exclusive right to subscribe to the increase of capital. The unlawfulness of this exclusion is a result of four reasons: - fraud to the law; - violation to the equality of the partners; - violation of the common interest of the partners; - impossibility to expropriate the reason for private utility."

The summons conclude: "Mr. Jean Loskron, who has kept his position as CEO, declares to the newspaper 'Ertak' of September 15, 1994: 'Loskron will be breakeven this year and will have a net profit of 5 percent in 1996'. For evaluation of their harm, the plaintiffs solicit an expert's report in order to assess the actualized value due from the future profits of Loskron. It is necessary to add to the material damage the intangible damage from the unlawful exclusion: the plaintiffs have signed with the Loskron company, in which they have agreed to invest, a contract that the company has violated in excluding them and depriving them from any right to participate in the turnaround of the company. In expectation of the final decision on their damages, the plaintiffs request the allocation of a reasonable provision of 1,000,000 F for all the plaintiffs. Finally, it would be justified to allocate for the plaintiffs the non recurring charges of the lawsuit amounting to 50,000 F."

The subject of this book is business ethics. The minority shareholders could have prevented the despoiling of their rights if they would have real power to oppose the resolutions of the majority shareholders. At least, they could have conducted a campaign in the press against Loskron, instigate a customers strike, contact the shareholders of Epsoks in London and notify them of the wrongdoing of Loskron, and so on. But ADAM has a mandate to restore the rights of minority shareholders in legal methods, although it is very difficult to win in those methods, as we shall see in this case. The summons was served before the tribunal de commerce in the region where the Loskron company operates, the address of the bailiff is at the Quai Loskron, in the name of the founders family of the Loskron company, the compensation requested is only one million francs, a minimal sum to justify a lawsuit of this scope, and most importantly – the plaintiffs are individuals with limited resources, including ADAM which consists only of its president, with an outstanding personality and exemplary courage, but who does not have a strong organization to support her. On the other hand, the defendants are a company that sells annually more than half a billion francs, they are the French leaders in their product line, and are backed up by one of the world leaders in this domain. This book will not deal with the legal ramifications of the lawsuit, but will follow it in its broad lines in order to try drawing conclusions from the case.

The plaintiffs request to compel Loskron to disclose if the bank debts were guaranteed by former majority shareholders of Loskron in 1993 and 1994, and if those guarantees are still in force or if on the contrary they were cancelled following the April and July 1994 protocols and/or the vote of 8.8.94. The minority shareholders try therefore to raise the curtain on the events that have preceded the vote of 8.8.94 and to verify if the majority shareholders are really so altruistic by having in mind only the interests of their employees as motive or if the reason that they agreed to sell the company was that it exempted them from honoring guarantees of hundreds of millions of francs.

"It is worthwhile to notice that, by the fact that the equity of Loskron was held by 75 percent by the Loskron family, the minority shareholders did not have any chance to make themselves heard in the vote. They were only given the possibility of selling their shares at a unit price of one franc, compared to a last quoted price of 64.90 francs." (L'Adam reclame l'indemnisation des minoritaires de Loskron, Adam requests the indemnification of the minority shareholders of Loskron, Trolley, 30.11.94) Apparently, the minority shareholders who held less than a quarter of the shares (the majority shareholders of Loskron held 87.8 percent of the shares following the offer to purchase the shares at one franc) did not have any possibility to prevent the sale of Loskron, and had only two alternatives: to lose all their investment or receive one franc per share, while the last quote was of 65 francs… About ten of the shareholders holding 4% of the equity of Loskron have therefore complained, together with Adam, before the tribunal de commerce in order to obtain damages because of unlawful expropriation at the purchase of Loskron by Epsoks.

The minority shareholders thought that after having lost all their investment, they did not risk anything in suing Loskron that had wronged them. They were mistaken, they lose the lawsuit, are treated as speculators and have to pay 410,000 F for abuse of rights. Henceforth, a norm is established, that minority shareholders risk more than their investment and the lawsuit expenses if they are wronged and sue the company. They have also to pay damages to those who have wronged them, if they lose the case. They are perceived as speculators, greedy to receive prompt and riskless benefits, while the majority shareholders want only the good of the company and safeguard employment. The minority shareholders are scorned, they are condemned, they are penalized. This is the result of an attempt to restore their rights, to sue the mighty companies before a court that possibly sympathizes and justify the actions of the majority shareholders. In the future, the minority shareholders will have serious fears and doubts before suing large companies. This example is very important to understand what were the motives of the minority shareholders in the following cases of Furolias and Mastoss, who have decided not to sue the companies, and those of Erinsar and Soktow who have decided to sue them. The minority shareholders sue the companies only in extreme cases or when they are backed up by large institutions or partners. On the other hand, if large institutions sue the companies, the latter try to accommodate them and find a compromise before the trial, as the companies are inflexible only toward the weak parties, but are very cooperative when they incur risks in confrontations with large organizations.

"While it wanted only to establish a principle on the undeniable rights of the shareholders, ADAM and other shareholders who applied with it to the court, found themselves seriously condemned by the tribunal de commerce of Lirotts under the pretext that they acted as speculators. We have to notice first of all that speculation is not unlawful in France and that there is no legal objection in this respect. Secondly, the plaintiffs wanted only to appoint an expert to evaluate the harm suffered from the suppression of the preferred right for subscription in the increase of capital incurred after the sale of ailing Loskron. It has to be known, first of all, that the value of the shares was brought down to zero. But even in such circumstance, the shareholders have to be respected: the possibility to participate in the reconstruction of the capital has to be preserved, in order, if they wish so, to be able to stay with the company. Too often, a company in bankruptcy could be a good bargain for a buyer who does not wish to share his good fortune with the other shareholders. Following that, they are indirectly expropriated, if the subscription right is eliminated.

It is on this basic question of principle that ADAM and 25 other shareholders have sued. They could have thought that they were on solid legal ground. The COB has often confirmed the preferred rights for subscription as an essential element of the property right. Unfortunately, in this case, the control body of the markets has given its consent to the increase of capital of Loskron without noticing the serious harm to the principles which it has itself formulated, henceforth, the impossibility of the COB to come to the rescue of the plaintiffs. Another delicate matter: it is reproached to the shareholders not to have asked the annulment of the extraordinary shareholders assembly that has decided on the increase of the capital. Truly, they did not want to act at this moment, fearing to risk that the purchase of the company will not be consummated and consequently harming the employees. Those despicable speculators, all so eager to maintain employment, are well compensated for their scruples. This fact does not prevent the court to speak of 'a concern of speculation performed to the detriment of a company whose primary concern is to avoid the disappearance of employment.' Thus, people who have lost the whole amount of their investment in shares of Loskron are furthermore condemned in common by a judgment applicable immediately (a respite is difficult to get and the appeal before the cour d'appel does not prevent the execution) to pay 410,000 F as damages for abuse of rights. The amount of those damages is dramatic for ADAM, which has to pay half of it. This association will find itself bankrupt and its president, Colette Neuville, can be forced to pay from her private assets." (Affaire Loskron: le droit de souscription bafoue, The Loskron affair, the subscription right is scorned, Insbol, 15.7.95)

ADAM was surrounded by benevolence and sympathy after this judgment and has received donations in order to be able to pay the considerable amount that menaced to sink the association. Insbol launches a campaign to assist ADAM and calls its readers to help: "After its condemnation, the survival of ADAM is threatened: the judgment is applicable immediately and ADAM has to get 200,000 F. Since the Tuffier affair, this association and its president Colette Neuville launch a tireless struggle for the recognition of the rights of the minority shareholders in France. Its disappearance would give a hard blow." In an edifying editorial the newspaper writes: "The decision that was taken by the tribunal de commerce of Lirotts in the Loskron affair, that we have mentioned in our last issue, is incredible. Not only the shareholders of Loskron, individually or represented by ADAM, did not win their case, but they were condemned, for abuse of rights, to pay 410,000 F. The tribunal de commerce of Lirotts has obviously its reasons. Justice is sovereign and, subsequently, its decisions have to be obeyed. In this particular case, they put cruelly in light the difficult connections between judges and shareholders. The first ones have always considered suspiciously the second ones. Their mistrust is a postulate. It suffices, for being convinced, to resort to the annexes of the annual reports of the Commission des operations de Bourse, where are reported the cases which have been brought up to justice: it is very rare that the minority shareholders plaintiffs have won their cases.

Another blatant proof: since Colette Neuville has started her struggle on behalf of the minority shareholders, she has won only one important victory, her first one: receiving compensation for the clients of Tuffier. However, this was a political-mediatic fight, but not a legal one. In her other fights, she has never won before the judges. Why? The reasons are multiple. Let us cite two fundamental ones. The first one, the most evident, is that for the judges, the shareholder is first of all a speculator, one who 'earns money while sleeping'. The first role of the shareholder, which consists in assisting the company in its development and to supply it the means, is not written down at all in the collective conscience. And even less in the judges' conscience. The second reason is no less serious: the shareholders are confronted with the incompetence of the judges. Let us be clear: the trade or business courts called tribunal de commerce are managed by local dignitaries who have never received formal financial training, while the subjects addressed are sometimes very intricate. From that emanates a reject phenomenon. The minority shareholders must therefore give up? 'No, says Colette Neuville, as our lawsuits do indeed make the law progress. Those are battering rams that shatter the doors and, with time, will force them to open.' If ADAM has lost in court against Pinault Printemps, this has nevertheless made the legislation progress over the public offers, in the sense of a larger equality toward the minority shareholders. Those actions have enabled some progress on the subject of mergers, or on that of expropriation of the shareholders. But if justice not only does not judge in favor of the minority shareholders, but penalizes them severely, the quest for equity begins to be very expensive." (Juges contre actionnaires, Judges against shareholders, Insbol, 22.7.95)

It is hard to add a better argumentation to this brilliant editorial, which summarizes all the dilemma of the safeguard of the rights of the minority shareholders by the courts. The minority shareholders who have been wronged cannot find consolation by the fact that the doors will be forced to open in the future. The methods of ADAM, the theses on ethics, and the media campaigns are important in the long run but do not give immediate results. As long as the judges, the companies, and most of all - society will continue to perceive the profits of the minority shareholders as the quarry of speculators and not as the milk of the poor's ewe, Nathan and the other ethical prophets will continue to preach in the desert. The best way open for the minority shareholders in the short term is probably to invest in ethical funds, in companies with an impeccable reputation of ethics, or to abstain altogether from investing.

This book, which is not a book on law, will abstain from raising the arguments of the Loskron company, the tribunal de commerce of Lirotts, the counter-arguments of ADAM, etc. which are very convincing, as everyone has his own truth, and the arguments for the reader, who is not himself a scholar in finance, can even be perceived as unwavering. All the appeals of the minority shareholders and ADAM have not succeeded to restore their rights and receive some compensation from Loskron, but nevertheless after the judgment of the Cour d'appel of 2.12.98, they were exempted from paying the fines sentenced to them by the tribunal de commerce of Lirotts. ADAM and the minority shareholders appealed to the Cour de cassation and are awaiting its verdict. But, miraculously, the OPA of Jean Loskron was made only a few days after the verdict of the Cour d'appel that has abstained from condemning Loskron. Could it be that Jean Loskron did not want this court to learn about the OPA to buy back his company, that could have shown a premeditation conceived since 1994 and postponed by the lawsuit of ADAM? Without this lawsuit, it is possible that Loskron would have been bought back by Jean Loskron much earlier, but that it was impossible for him to do so in order not to raise the suspicion of the Cour d'appel, which could have asked how is it possible that a company that was almost bankrupt and bought by an Anglo-American group is being bought back by the majority shareholders of Loskron, that pretend at court that they have lost all their investment.

This book is, as mentioned before, a book on ethics that seeks to prove that the minority shareholders were wronged ethically and not legally. This is the difference between ethics and the law. The case has therefore proved that the ethical conduct of Loskron toward the minority shareholders has undoubtedly despoiled them, especially within a five-year retrospective. The only partners who were wronged are therefore the minority shareholders, who have lost all their investment and did not have the opportunity to win their case through the assistance of the French legal system!

9

CASE STUDY OF THE ISRAELI/AMERICAN COMPANY FUROLIAS

"Obsequium amicos veritas odium parit"
Readiness to oblige creates friends, frankness engenders hatred
(Terence, Andrienne, I,1,68)

"While it is true that many criminals do tend to ignore their victims and/or to blame their victims; and while it is plausible to associate these tendencies with a penchant for doing acts that promise immediate gratification and little risk of being caught, there are other features of evil that are not so plausibly associated with low self-control. Consider the following features of evil (or evil agency) discussed by theologians, philosophers, novelists and other thinkers who are not inclined to romanticize evil:

1 Evil acts are mechanical or repetitive. The agent might be described as an automaton (Ouspensky, 1949).

2 Evil acts arise out of identification with a group. Fidelity to the group results in agents who are willing to commit all sorts of atrocities in the name of protecting the group (Unsworth, 1982).

3 Evil involves ignoring an inner voice that warns against some act (Unsworth, 1982; Plato, trans. 1971; Thompson, 1952) and/or external voices that counsel against performing the action (Thompson, 1952; Lewis, 1992).

4 The evil agent construes himself as a largely passive being who is under some compulsion to perform an act which others subsequently judge evil. This agent may do acts because of the appearance of some 'omen' or 'sign' (Thompson, 1952).

5 Evil acts are done by agents operating in a dreamlike state (Unsworth, 1982; Thomson, 1952; Ouspensky, 1949).

6 Evil acts are done out of a belief that anything is possible (Sereny, 1995; Buber, 1953; Arendt, 1979).

7 Evil stems from the agent's certainty that his or her action is absolutely justified (Plato, trans. 1971; James, 1986).

8 The agent who has done what is deemed evil uses the language of morality to justify the immorality (Sereny, 1995; Aristotle, trans. 1975).

9 Evil involves an element of bad faith (Sartre, 1993).

10 The evil person attributes to another person or group exactly the deed he himself is about to perform, a deed which victimizes this person or group (Arendt, 1979; Young-Bruehl, 1996).

11 Evil agents cast themselves in the role of a benefactor. The evil person demands 'purity' from others and then promises continued beneficence if only the other party will remain pure. The continued beneficence is often made contingent upon the party's willingness to openly 'confess' any and all past impurities (Lifton, 1989; James, 1984)."

(Business Ethics Quarterly, January 1998, Koehn, Employee Vice, p. 152-3).

It was the eve of Yom Kippur, the Day of Atonement, October 6, 1992. Ulysses Karisios went to visit his mentor and friend Gorekius to wish him a happy new Jewish year and the forgiveness by God of all the sins of the previous year, as is the custom in Israel and all over the Jewish world. Karisios, in his late forties, spent most of his business career working at Erinsar, one of the largest high-tech companies in Israel, founded by Gorekius, who was 20 years older than Karisios. Karisios reached top-level positions and was responsible for the business side of the company, which became the most profitable company in Israel. He reported to the President of Erinsar – Istovius, who became a personal friend over the years. The morale at Erinsar was very high, the management operated in harmony and most of the Vice Presidents were also friends, adhered to each other and to the company and shared the success of Erinsar. However, in 1987, Karisios noticed a change in the ethical conduct of Istovius. In order to remain a leader in its business, Istovius navigated the company in a course that contradicted fundamentally Karisios' ethical beliefs. When he could no longer bridge between business and ethics, Karisios resigned, although Gorekius, the Chairman of the Board of Erinsar, tried to dissuade him to no avail. However, the friendship between Karisios and Gorekius was not tarnished, as Karisios did not make public to the press his dissent and kept the disagreement within the 'family'. Karisios started a new career as a consultant and was very successful. Erinsar, managed by Istovius, continued also to succeed, although the ethics deteriorated more and more and many members of the top management left the company.

Karisios was, among others, a consultant of Furolias, a medium-size company founded also by Gorekius, in a business line distinctly different than those of Erinsar. Furolias' high-tech business amounted to $54.3M in 1992 with a loss of $25.2M. Furolias succeeded in reducing its losses in 1993, after implementing a turnaround plan, to $6.2M with sales of $47.6M. Karisios started to work with Furolias in 1991 on a small scale and over the years Furolias became Karisios' main client, and although the company encountered heavy losses he was confident that it had a bright future. Karisios invested most of his savings in shares of Furolias, which traded at NASDAQ at an average price of $6, about half of the price that it traded when the company was profitable. The management of the company was confident that the new products that Furolias has developed had a tremendous potential and was very optimistic about the future of the company. This belief was conveyed in private conversations, in meetings with the Government authorities, customers, banks, and also in articles in the major newspapers in Israel. Nalodo, the parent holding company founded and managed by Gorekius, stood behind Furolias and helped the company overcome its problems as it did in similar cases with other subsidiaries. The price of the shares was quite volatile due to a low trade, and the market valuation of the company averaged $50M, similar to its sales volume, although it never went about 20 percent below this valuation until 1994.

At the meeting between Gorekius and Karisios in 1992, Karisios mentioned to Gorekius for the first time that he was a consultant of Furolias, as he did not want to take advantage of his friendship with Gorekius in his relations with the management of Furolias. Gorekius was the Chairman of the Board of Furolias, of Erinsar, as well as of Nalodo, the parent company of both Furolias and Erinsar. Gorekius confided to Karisios that he wanted to gain full control of Furolias, in which Nalodo held only 40 percent but effectively controlled, but he could not do that at the existing price which was too high, although he was convinced that Furolias had a very bright future. He was very pleased to hear that Karisios was a consultant to the company and had even invested into it. Karisios, who strongly believed in the ethics of Gorekius, who had an impeccable reputation, did not guess what could be the outcome of a potential strategic takeover of Furolias by Nalodo's Group.

Karisios continued to work with Furolias for one more year and left the company in excellent relations with Wersnon, the President of Furolias, and continued to assist him on a friendly basis whenever Wersnon needed it. The shares' price of Furolias started to fall abruptly shortly after Karisios left the company. Karisios, who sensed that something was wrong with the conduct of the shares' price, could not sell the shares as he did not want to be accused of using insider information, although he no longer worked for the company. He knew that several companies were interested in investing in the company, as Gorekius asked him to assist him in finding an investor, and he contacted several companies on Gorekius' behalf. He decided to wait a few months as a cooling-off period to see what will be the outcome of those contacts, as he was still convinced of the future of the company. However, in those months he started to suspect that the fall of the price was possibly instigated by the parties involved. Dorian who worked with Istovius in Erinsar in 1992 and was transferred to Furolias told him how Istovius confided to him that Erinsar planned to takeover Furolias as early as 1992, the same year in which Gorekius confided to Karisios on the same subject. Apollo told Karisios how his equity fund, which was founded by Durtem, the bank that was also directly and indirectly the largest shareholder of Erinsar, Furolias and Nalodo, sold his shares in Furolias just before the fall of the shares' price. More and more evidence of unethical acts of Gorekius started to permeate and a premeditation between Nalodo and Erinsar became apparent to Karisios, although he did not have any proof of it. In the meantime, the shares' price collapsed to about 80 cents and on the first of July 1994 Erinsar tookover Furolias to the complete surprise of the business community in Israel.

More and more facts indicated that the merger resulted from a possible premeditation conceived a long time ago between Nalodo and Erinsar, with the cooperation of Furolias. A few months before the takeover, Tevel, the treasurer of Nalodo, asked Karisios why he does not sell his shares. Karisios answered that he was waiting for them to return to $6, to that she answered: 'this will never happen!' Karisios heard from three friends - Barad, Amir and Orion, who were involved in Gorekius' contacts with investors for selling Furolias, that Gorekius did everything to discourage them and emphasized all the risks involved in the purchase and the precarious financial condition of the company. The official excuse for the takeover was that Erinsar was the only company that made an offer to purchase the shares of Furolias for the current market price of about $1, thus acquiring for about $8M, payable in shares of Erinsar or 2 percent of Erinsar's shares, a company selling about 50 million dollars, that has managed to reduce its losses from $25M in 1992 to $6M in 1993.

Karisios' lawyer, Yraye, a friend and one of the most prominent lawyers in Israel, sent a letter to the CEO of Furolias, Wersnon, on July 12, 1994 asking 18 questions about the negotiations between Erinsar and Furolias, the involvement of the board of directors of Furolias, including the independent directors, the negotiations with other potential buyers, the promises that Wersnon might have for continuing to work for the group, etc. Immediately after sending the letter, Gorekius tried to meet Yraye without success. Hustash and other good friends of Karisios working with Erinsar, contacted Karisios and tried to figure out what he really wanted. On July 14, 1994, only two days after Yraye's letter, he received an answer from Robin, the lawyer of Furolias and a friend of Karisios, from the law firm Jonroms, one of the largest law firms in the US, who represented also Nalodo and Erinsar, but 'not in this transaction'. Furolias was officially a US company, although its headquarters was in Israel, and most of its legal matters were held by this US law firm. The letter was very short and polite, and stated: 'Please note that we are in the process of working with Furolias to prepare a detailed proxy statement, in accordance with the rules of the U.S. Securities and Exchange Commission, describing in detail the proposed merger and all material related facts. I believe that virtually all of your questions will be fully answered by the proxy statement.... We have been in direct contact with Furolias with regard to its disclosure and corporate obligations, and, as far as we are aware, Furolias has fully complied with all of its obligations under Delaware corporate law and the U.S. federal securities laws. To the extent that you have further questions of a legal nature, please direct these questions to our firm.' What Robin did not mention in his letter was that probably he was very thankful for Yraye's letter, as the questions helped Furolias to correct many loopholes that existed in the original merger agreement and that by the end of the year, when the merger was finally approved, the proxy documents were almost 'perfect'.

Nevertheless, Yraye's letter embarrassed Gorekius. Karisios was not just a minority shareholder, he was one of the 'family', a former VP of Erinsar, a senior consultant of Furolias, and a personal friend of most of the management of the group. Gorekius planned to make two offerings of Nalodo and Erinsar shares, based on the merger with Furolias, in August 1994, and a lawsuit against them would endanger the offerings. Gorekius was probably intrigued about whether Karisios had 'smoking gun evidence' for his allegations as Karisios knew everybody in the organizations and not all of them were 'fully reliable'. Gorekius did not succeed in learning much from the various conversations of Karisios' friends with Karisios, he did not know, unless he wired Karisios' phone, that Karisios has decided not to sue Furolias, without the backing of a large company. Karisios was afraid that if he sued them alone, he might encounter the risk of being sued by Furolias, as Erinsar has done in the past in other cases. Karisios remembered the case of Akteon, which occurred a few years earlier. Erinsar was sued by this US company for $20M prior to Erinsar's public offering. Erinsar decided not to compromise as they sensed that their opponent was not strong enough. They convinced the US court to dismiss the case and sued Akteon in Israel. In the US, Jonroms handled the case, the same company that represented Furolias, and in Israel, Bronf handled the case brilliantly, and succeeded in winning the case. Akteon was sentenced to pay Erinsar indemnification of $XM...

Nalodo and Erinsar decided probably to adopt a stalling tactic toward Karisios. Istovius contacted Karisios and spoke to him in a very friendly manner. In the past few years they barely spoke once or twice, due to the stormy resignation of Karisios from Erinsar. Istovius evoked the good old days and said that they should renew their contacts. He stated that he had nothing to do with the allegations of Karisios against Furolias and Nalodo, as this transaction was for Erinsar an occasional windfall. However, he was willing to discuss several alternatives that 'were not connected to Furolias' and that could give Karisios a compensation similar to what he has lost in Furolias. Karisios, who was flattered and allured, forgot what he learned bitterly in the past from his contacts with Istovius (and from the fable of La Fontaine) and fell into the trap. Istovius was all the time travelling abroad and could not meet Karisios, and in the meantime Nalodo and Furolias issued two successful public offerings based among others on the Furolias transaction and raised tens of millions of dollars. When Karisios realized that he was conned it was too late. He wrote a brilliant synopsis of a statement of claim/complaint to the SEC and on September 17, 1994, a Saturday night, he visited Otwuss, one of the leading lawyers in Israel, a Professor of Law, and a prominent ethicist in law and business. Otwuss, acted as the Israeli lawyer of Furolias in this case, and in other cases as the lawyer of the Durtem group, Nalodo and Erinsar. Otwuss was one of Karisios' best friends, they succeeded in the past to make together many successful deals, when working at Erinsar and as a consultant, they respected each other, and Karisios had no doubt about Otwuss' integrity and ethics.

Karisios showed Otwuss the synopsis, told him how he was conned and said that if he will not succeed in finding a partner for a lawsuit he will disclose all the affair to the SEC. The merger at this time was not completed. It was to be completed only three months later. Karisios was rather baffled by the potential dilemma of how Otwuss could reconcile his integrity with Furolias' acts, but if lawyers represent criminals why shouldn't they represent 'honorable companies and executives'. As far as their friendship was concerned, he had no illusions. In the harsh business atmosphere in Israel, as in the US and France, friendship has no place when interests are involved. However, he knew that only Otwuss had the moral weight to convince Istovius to fulfill his promises.

Otwuss was astonished from the content of the material and he asked for a 24-hour grace period. The following day Istovius contacted Karisios and they set a meeting for September 29. After reaching an agreement that fully compensated Karisios for his losses, yet not mentioning at all Furolias, they decided to sign it on October 3. However, an hour before the signature, Istovius phoned Karisios and said that he was sorry that he had mislead him but he had received a legal opinion that as Karisios is a 'dissident shareholder' they cannot sign the agreement. He suggested that they meet the following day with Gorekius and try to find a solution. The three of them met at lunch on October 4, 1994, but to no avail.

From the first days of July 1994 Karisios tried to approach the shareholders of Furolias in order to share the information for a potential lawsuit. He held less than 1 percent of the shares and he knew, as stated above, that alone, without resources, he would not be able to act effectively against a holding company controlling companies with sales of more than a billion dollars, owned by one of the largest banks in Israel, and backed by some of the richest men in Israel with assets of billions of dollars. He contacted Doherty, a US equity fund, that owned more than 5 percent of the shares. Furolias was registered in Delaware, traded at NASDAQ, thus operating as a US company, although most of its operations were in Israel, including its headquarters. He spoke with Noskivar, who reacted to what Furolias has done, saying: 'it smells a rat!'. However, two months later on September 16, 1994, Karisios learned that Doherty decided to vote for the merger. Of course, it was impossible to assess if Furolias had reached an agreement with Doherty, but in view of the initial reaction of the fund and the fact that Nalodo was always unwilling to confront strong opponents, Karisios could only conclude that Furolias/Nalodo/Erinsar had found a way to compensate Doherty. He was contacted by Barad who was willing to cooperate, but he was too small an investor. Prior to the takeover, Barad brought a US investor to purchase Furolias, but he said that Gorekius dissuaded him from doing so. However, as he was an employee of Furolias, he mentioned to Karisios many acts of the company that apparently were illegal, and was told by one of the Vice Presidents of Nalodo of an illegal conduct involving Gorekius personally. It was a good start, but still Karisios and Barad did not have enough funds to sue the companies and they needed a strong partner.

Karisios went to other shareholders, but found to his dismay that the rule of Omerta prevailed. One of them, the owner of one of the largest high-tech companies in Israel, even told him that he prefers to incur the loss than to start a lawsuit, as he doesn't think that lawsuits against colleagues are appropriate. Shortly afterwards the newspapers reported that his company and Furolias were going to collaborate in the development of a new product. Orion, a European distributor of Furolias, who lost more than a million dollars in his investment in Furolias' shares, was in particular angry at Gorekius, as he wanted to purchase Furolias and Gorekius dissuaded him to do so. However, when the chips were down, Orion backed off as well, as he did not want to lose the distributorship of Furolias in his country.

Amir, a colleague who told Karisios that his investor for the purchase of Furolias was dissuaded by Gorekius from purchasing the company could not help Karisios as he was married to the daughter of the President of the parent company of Nalodo… Cornfeld, the President of an investment subsidiary of Durtem, the owner of Nalodo, who wanted also to purchase Furolias was friendly persuaded to back off. Many other friends gave Karisios all the information needed to be convinced of the premeditation of the takeover at the detriment of the minority shareholders, but none of them was willing to testify as they did not or could not risk their job for a friend in need. Some of them also asked Karisios: 'would you act in my place differently?' Hartishna, a US friend who was also the lawyer of Nalodo and Erinsar, told Karisios to back off as his opponents were very mighty and he could lose all his clients who wouldn't want to deal with a troublemaker. He risked to be treated as an 'untermensch', a Yiddish word for subhuman. Karisios commented that he had time and would take his revenge as Sadat did in the Yom Kippur War. To that she answered: 'Look what happened to him...'

When Gorekius approached Karisios personally, he tried to convince him that all that was done was completely legal, Erinsar did not want at all to purchase Furolias and he had to convince Istovius with great difficulty to purchase the company as otherwise hundreds of employees would have found themselves without work. We remember that that was also the argument of Jean Loskron in the takeover of his company Loskron, as he only wanted to save the jobs of his employees. Furthermore, Bsosskins, one of Karisios' best friends who remained at Erinsar and managed the takeover, told Karisios: 'Why are you complaining? I remember that you used to speculate when you were at Erinsar. You win some and you lose some.' Here again, the small investor is called a speculator, exactly as the French Court called the poor investors of Loskron, who lost all their money. In Bsosskins' case it was more than cynical, as Karisios not only did not speculate at all, but almost all his savings were invested while working in Erinsar in Erinsar's shares, which he refrained to sell when they were at their peak, as did most of his colleagues, but at a discount of 30 percent to 40 percent, as he did not wanted to be accused, as the CFO of Erinsar, of using insider information. Gorekius also argued: 'You are the only one to complain. Some of my best friends lost a lot of money in my companies but none of them complained!'

Before appealing to outside parties, Karisios tried a last move. He went to Diarkra, a good friend and a Vice President in the holding group of the Durtem bank, Nalodo, Erinsar, and all the other companies involved. He gave her the material and asked her to show it to the 'ruling family'. A few days later she sent him back the material and said that it had been reviewed and found serious but that they would not interfere in the acts of public companies that had their own board of directors.

One of the most characteristic differences in comparing the cases of Loskron in France and Furolias in Israel that happened exactly at the same time is the coverage of the press. While in France most of the coverage criticized the deal and emphasized the wrongdoing of Loskron, mainly because of the interference of Mrs. Neuville, the President of ADAM, the Israeli press was almost unanimously sympathetic to the deal. The main reason was of course that nobody told them of the severe implications to the unaffiliated shareholders, as it was done in the next case of Erinsar and Soktow. But also, Erinsar and Nalodo have invested over the years substantial amounts in PR and maintained excellent relations with the press. The Israeli press is often accused that because they are owned by some of the richest families in Israel who maintain excellent business contacts with the other '50 ruling families', and because they are very dependent on the advertising budgets of the largest corporations, they prefer not to disclose scandals involving prominent businessmen, unless there is 'smoking gun evidence' that could convince and be understood by the average reader.

The coverage of the newspapers reflected therefore mainly what was told to them by the management of the group. "The management of Furolias expects that the company will return to profitability in 1995. Erinsar announced yesterday the completion of the merger with Furolias, and published a proposal to purchase the shares of Furolias from their shareholders. Erinsar proposes to the shareholders of Furolias one share of Erinsar for 20 shares of Furolias. Upon completion of the merger, Furolias' shares will cease to be traded on NASDAQ in the US. Furolias is a failing company in... The company lost in the first six months of 1994 $7.2M, with sales of $15.9M. It is estimated that the company was expected to crumble completely within a year, unless it merged with the affiliated company Erinsar. The company moved recently to a new product line, and needed an investment in order to finish the development and manufacturing system of the products. The merger decision was reached, after no other investor or strategic partner was found for Furolias. The President of Erinsar, Istovius, estimates that the merger between the two companies will assist Furolias to benefit from its technological and sales potential." (After the merger with Erinsar: Furolias is expecting profits, Yarmuk, 8.8.94)

The reader of this article understands that a failing company with a life expectancy of a few months and no other investors willing to invest in it, was rescued by an affiliated company in order to benefit from its potential. The shareholders of the company are probably offered a fair deal, because their shares will no longer be traded over the counter. Only after reading the whole case in this book and understanding the intricacies of the transaction one could learn that there were other investors who were discouraged by the Group, that the company could survive and finish its projects as almost all the investments were already made and Nalodo promised to back up Furolias, that the shareholders of Furolias could have been offered a rights issue and could still sell their shares, and that the company was not at all failing, as the losses in 1994 included reserves, that one of its senior managers was willing to testify that they were doubtful.

Strikingly, Furolias' arguments, the return to profitability expected in 1995, the doubtful reserves, the failing impression conveyed by the company, the lack of other alternatives, and so on resemble exactly the Loskron case. If the cases did not happen simultaneously in France and Israel, with shareholders completely different and in unconnected business lines, one would tend to think that the tactics were devised by the same people. But probably great minds think alike, and there are inflexible rules that apply to many similar cases as will be shown in the next two cases.

One could ask: 'Why didn't Karisios contact the newspapers as Mme. Neuville did?' But in this period Karisios was still trying to find an amicable solution within the family and, anyhow, we see that the results of the press campaign of Mme. Neuville in the Loskron case and of Astossg in the Erinsar case were nil. Furthermore, when Karisios ultimately approached one of the leading newspapers they did nothing. Press campaigns in such intricate issues are probably not the right method to safeguard the minority shareholders' interests.

All these arguments and behavior prove the accuracy of the article on Employee Vice, quoted at the beginning of this case. The 'heroes' of this case, who were Karisios' friends, All K's Friends as in All My Sons of Arthur Miller, acted, as Karisios believed, in an evil manner out of identification with a group. We remember that at Erinsar, there was a strong feeling of adherence. This feeling, which was channeled at the beginning to a constructive path, deteriorated, according to Karisios, during the years and throughout the loss of ethics to an evil path that based the growth of the company on obstructing the rights of the partners, the clients, the government, the employees and of course – the minority shareholders. They ignored their inner voice and their basic honesty, as they wanted to succeed at all cost and as they did not manage to do so on their merits, due to the fact that in many cases the less talented senior managers remained with the company. They succeeded mostly by trespassing the rights of others, as this case and the Erinsar case will prove later on.

Karisios was convinced that they believed that anything was possible, as they were not punished by their behavior and they were backed by the richest men in Israel. They thought that their action was absolutely justified, as there was no other alternative, and minority shareholders, just as in other regimes - minorities in general - were 'expendable'. They used the language of morality to justify their immorality, in extremely bad faith, attributing to other persons exactly the deed they performed, thus calling the shareholders speculators. The evil managers believed without reserve in the company they worked with and that was their method to assuage their conscience, if a vestige of it still remained in them. Most of the arguments of the article are therefore evident in this case.

In the meantime, Karisios was convinced that Gorekius used insider information not to invest in Furolias and to invest at exactly the same period in Memnit. The same amount invested by Karisios in Furolias was invested at the same period by Gorekius in Memnit, and instead of losing 90 percent of his investment, Gorekius managed to earn more than $10M. Karisios learned about Gorekius' illegal action from Barad, but could not substantiate legally his claim, as the witnesses would not or could not cooperate in the lawsuit. Karisios was most of all annoyed by the fact that evil agents cast themselves in the role of benefactors. They receive Doctorates Honoris Causa from the best universities, the highest honors and prizes, are called the founders of the high-tech industry, and benefit from having the best reputations. All that, while the offended parties are called speculators, misfits, troublemakers, dissidents, whistle-blowers and subhumans.

Ultimately, the 'troublemakers' can only cause troubles that annoy the large corporations and their mighty shareholders like a flea bite on an elephant. Karisios told his story to all his friends, or what was left of them, but this did not tarnish the reputation of the management and owners of the Group. 'One cannot argue with success' was the answer that Karisios received from his friends. On the contrary, Karisios' conduct affected his reputation, as he had evidence in writing that his opponents spread slanders that he was dismissed shamefully from Erinsar, which was untrue as could be proved by him with all the documents that preceded his resignation and by the fact that he continued to work as a consultant for the Nalodo group after he left Erinsar. Karisios could sue them for slander and receive a judgment after five years and enormous costs, but he preferred no to do it.

Karisios retaliated as best as he could. As the Group was aware of the fact that Karisios knew of many unlawful acts, they had to change during the six months of the 'negotiations' between Erinsar and Furolias, many plans that cost them after all several millions of dollars. Wersnon, the CEO of Furolias, who was perceived as an independent director, but in fact Karisios was convinced that he cooperated fully with the Group, could not receive a job within the organization as it would prove the motives of his collaboration. Whenever Wersnon applied for a job and Karisios learned about it, he called the owners of the companies, whom he knew and told them the whole story, thus stalling Wersnon's search for a new position. But of course, all this changed effectively nothing. Still there were some friends, such as Nisan and Zivav, who assisted him in his need, arranged for meetings and contacts with influential people, reviewed and commented his claims and were not afraid of the negative implications it could cause them, should it be known to the Durtem Group.

Karisios came to the conclusion that the only way to fight this evil is through ethics, as the legal system cannot assist in most cases the weak parties, after his attempts to resort to legal suits proved unsuccessful. Karisios approached Wasniss, one of the most prominent lawyers in Israel, who had won a lawsuit against Erinsar in the past. Wasniss read the material and said it was 'explosive', but he insisted that Karisios would find a strong partner for his suit. Karisios sent also the material to a prominent law firm in the US that specializes in class actions. They agreed to handle the case but asked for such upfront fees that were prohibitive for Karisios alone.

After many inquiries, Karisios discovered that Lupinus, an equity fund of Boral, one of the largest banks in Israel, had invested in Furolias and lost a large amount of money. He met Zahav, the President of the fund, on November 22, 1994 and urged him to fight this case. Zahav decided to send Umberto, one of his company's executives, to the shareholders' meeting of Furolias that was held in New York on December 30 (!) 1994, (we remember the shareholders' meeting of the French Loskron that was also held in August 1994, the month of the French vacations) in order to ask Furolias' management 20 questions that were raised by Karisios. However, Umberto was answered evasive answers and the questions and answers did not even appear in the report of the meeting sent to the shareholders. Furolias even prohibited Zahav and Umberto to show Karisios the protocols of the questions and answers...

The takeover of Furolias by Erinsar was approved by an overwhelming majority, 5,175,000 votes in favor and 200,835 votes against. But effectively Furolias was already managed by Erinsar since July 1994, based on a management agreement signed by both parties. The shares continued to be traded until June 1995 and the last trade was at $7/8. Karisios and all the other shareholders had no other alternative but to receive Erinsar's shares. Karisios retained one Furolias share for a potential claim and sold the Erinsar's shares that he has received on the same day, losing about 85 percent of his initial investment in Furolias.

Zahav and Karisios tried to organize the minority shareholders against Furolias, but the US legal advisors of Zahav found out that it was illegal to do so according to the Delaware corporate law... They requested the list of the 700 shareholders of Furolias, but Furolias refused to comply. When Karisios sensed that Zahav's firmness was floundering, he approached a member of the Board of Directors of the fund. Karisios sent official letters to Zahav and continued to raise the issue at the Board of Directors of the fund several times. But on October 23, 1995, Zahav decided also to back off. After all, it is probably very difficult to resist pressures, even when you have the best intentions, specially if the fund is managed by the Boral bank group, which has also substantial equity in Erinsar and Nalodo, and has deposits of tens of millions dollars of Erinsar.

Karisios did not despair, on November 9, 1995, he met Eikon, the auditor of Furolias, Nalodo and Erinsar, and a personal friend, who was present at the Furolias' shareholders meeting in New York in December 1994, and Karisios gave him all the material he had against the Group. Eikon, whose firm Lortow was amalgamated recently with Ascorage, one of the big six, was very preoccupied and tried to dissuade Karisios from giving him the material, but eventually, after receiving reluctantly the material, he did nothing. Prior to that Karisios also gave the material to the Israeli SEC through Zahav on December 28, 1994, and finally he sent all the material to the US SEC and met on June 25, 1996 with a reporter of one of the leading Israeli newspapers. All of this without any success.

Nobody was interested in investigating the case, there were no solid legal proofs, 'it was not in the public interest', in short Karisios, the minority shareholder, did not have the weight to start an investigation. A renowned private investigator proposed in a meeting held on September 5, 1996 to conduct an investigation on the possibility that Erinsar and Nalodo manipulated the prices of Furolias' shares, discouraged on purpose Furolias' potential investors, premeditated the takeover of Furolias by Erinsar prior to 1994, and conducted fraudulently toward the unaffiliated shareholders of Furolias. He asked for a huge amount for conducting the investigation but Karisios' partners were unwilling to invest, and it was beyond Karisios' means to do so on his own.

The most intriguing conduct in this case was that of the Israeli and US SEC. The auditors' attitude may be excusable, as they did not want to lose one of their main clients, although their mandate is to make full disclosure of the companies' data. The management of the companies do not want to lose their jobs, the board of directors are appointed by the majority shareholders and the independent shareholders in most of the cases do not understand or do not want to understand what it is all about. The lawyers who defend also thieves and killers are convinced that their mandate is to defend their clients, especially if they are the largest companies in Israel. The friends risk their friendship, which is a small price compared to their relations and their reputation as conformists.

Only the SEC is supposed to be the safeguard of the minority, yet when it was approached by Zahav, one of the top officers of the Israeli SEC answered him after reviewing the material that they could not interfere as Furolias was a US company. So much US that most of its operations were held in Israel, the headquarters was in Israel, and that the Israel citizens were allowed to invest in Furolias and receive the same tax benefits as did all the other Israeli companies. Furthermore, two offerings that were issued in August 1994 by Erinsar and Nalodo, two Israeli companies, were based on the merger with Furolias, and contained all the issues that were treated by the documents shown to the SEC by Zahav. In order to retain an impeccable reputation, the Israeli SEC could also forbid its officers to be hired after retiring by the companies they had supervised, as this was the case with one of the leading managers of the SEC who was appointed to one of the top positions of the Durtem group, the parent company of Nalodo, after the events of this case took place.

So, if Furolias is a US company, the only way that remains is to complain to the US SEC. Tirelessly, after talking to Sorial on the phone on July 10, Karisios sent on July 11, 1996 a letter to Sorial of the Enforcement Department of the SEC. He learned about Sorial following a recommendation of Shiran, an Israeli governmental legal counsel who volunteered to help. This letter, which could no more assist the minority shareholders of Furolias, was sent in anticipation of the wrongdoing to Erinsar and Soktow's shareholders that Karisios was convinced would happen and that will be treated in the next case. The writing was on the wall, the pattern was obvious, and if the US SEC would have investigated Furolias' case on time, the Erinsar and Soktow shareholders would not have lost hundreds of millions of dollars. We remember the article quoted at the beginning of this case that states 'Evil acts are done out of a belief that everything is possible (Sereny, 1995, Buber, 1953, Arendt, 1979) After succeeding in depriving the rights of the minority shareholders of Furolias in tens of millions of dollars, Erinsar and Nalodo were apparently ready to reach the next stage of hundreds of millions of dollars. But in a conversation between Karisios and Sorial on September 25, 1996, Sorial told him that the SEC decided not to handle the case, as the document sent by Karisios had no evidence requiring an investigation by the SEC.

As this is the crux of the matter, this case gives in full the letters sent to the SEC and the synopsis of the events sent to the SEC, as follows:

"Following our conversation and the recommendation of Mrs. Shiran, we hereby submit for the SEC's investigation a claim against Furolias, Erinsar and Nalodo, all of them traded at NASDAQ, that was originally intended for a class action. Due to the high costs of legal proceedings in the US, we resort to your assistance in order to find justice and to prevent the occurrence of similar wrongdoing to other unaffiliated shareholders of those companies. We tend to believe that the imminent split of Erinsar into three companies and a potential share purchase offer to the unaffiliated shareholders of Soktow, a subsidiary of Erinsar, may result in similar extortion.

We are sending you a Synopsis of Statement of Claim by the Unaffiliated Stockholders of Furolias. Furolias International, Inc., a ... products company, traded over the counter as Furolias..., has ceased to exist in January 1995 and all the holders of its shares received 1/20 Erinsar's share for every Furolias' share. Erinsar (NASDAQ – E....) is a one billion dollars conglomerate and one of the leading Israeli companies in ... Nalodo (NASDAQ - ...N.) was the holding company of Erinsar and Furolias and had about 40% of each company's shares. 58% of Furolias' shares were held by unaffiliated shareholders and the majority of them did not vote for the merger.

We are a group of 3 shareholders of Furolias that held prior to the merger 2% of its shares, but many other shareholders sympathize with our endeavor. I (Karisios), personally worked as a consultant for Furolias during 1991 - 1993 and as a VP of Erinsar in 1981 - 1987, was a close friend of Mr. Gorekius, the Chairman of the Board of Nalodo, Erinsar and Furolias, until the merger,... I have invested heavily in Furolias, believed in its potential and did not sell my shares even when the company lost $25M, two years prior to the sudden and dubious collapse of its shares. I was also very cautious, in order not to be blamed to take advantage of any insider information, as it was public knowledge that I held very close contacts with the management of the company and its Chairman, who knew of my investment.

The synopsis is based only on evidence that is known to the public and was published in official statements and in the media. We believe that the issues that are referred in page 23 of the synopsis can be backed by confidential information. We believe that some of the major US unaffiliated stockholders were offered a compensation in order not to sue Furolias. We know that prior to the collapse of Furolias' shares, Irkuson, a Durtem Group mutual fund, sold all of its Furolias' shares. The Durtem Group is the largest shareholder of Nalodo. I have personally warned the parent company of the Durtem Group, and Eikon, the CPA of Erinsar, Nalodo and Furolias, of the wrongdoing of those companies, and gave them the relevant material, but to no avail.

After the unfriendly takeover of Furolias by Erinsar, we have made our own inquiries on the integrity of the Board members of Furolias. One of us heard personally from Posturck, a member of Furolias' Board, that he was sent by Nalodo to conduct a due diligence on Memnit in 1991, prior to Nalodo's and Gorekius' investment into the company. Posturck told him that Gorekius gave him at the same time a personal loan of $20K to invest into the company, with no obligation to return the loan, should the investment prove not to be successful. Posturck boasted that this risk free investment turned to be the most successful investment of his life, as its value increased to $4M. Gorekius and Posturck made full disclosure of their investment, but forgot to mention the risk free loan. Gorekius last invested into Furolias in 1984 and Posturck never invested into the company. This is only one example, but we have gathered many other informations that could be of interest to you about Erinsar and Soktow as well.

We are confident that your investigation of the companies' executives and directors, especially of those that are no more part of the Durtem Group, will reveal the truth about Furolias and the other companies of the Group, as some of these persons have already expressed their dissent from those companies' mode of operations. This could only be done by the SEC or by the Israeli press, as many of the people involved are afraid to disclose all what they know, due to the fact that, unfortunately, we are still ruled by the laws of Omerta. We could disclose to you evidence about the threats that we have encountered, including in conversations with the American lawyer of the companies. The Israeli SEC was approached by Lupinus, a mutual fund of Boral, that has also invested heavily in Furolias, but they said that as Furolias is formally a US company registered in Delaware, they cannot investigate the case.

We would appreciate your prompt response and are willing to give you all the informations required. We could meet in New York or in Washington on August 1 or 2. Please, do not hesitate to contact me by phone ... , by fax ... , to my mobile phone ..., or to my home address ... .

We thank you for your cooperation and do hope that even if we have lost almost all of our investment, justice will be done and other shareholders of Nalodo, Erinsar and their subsidiaries will not encounter the same fate."

Karisios phoned Sorial several times to inquire about the potential investigation. He even sent him another letter, which was material to the case, on August 28, 1996.

"Following our letter and document from July 11, 1996 on Furolias, Erinsar, Nalodo and the group, I hereby send you an article published today by the largest Israeli newspaper, Yarmuk, about the arrest of Zupon, the former general manager of the Toren and Furolias division at Erinsar.

Zupon who was arrested and questioned for 48 hours, is suspected of giving instructions during the years 1992-1996 to present false cost documents to the Israeli customs authorities, as a basis for the reports of the Toren sales taxes by Erinsar, which is suspected of evading taxes of 20 million shekels, about $6M.

Erinsar denied the allegations. Zupon is the third Erinsar officer arrested in this context. He was released on bail. Zupon is no longer working at Erinsar.

We believe that Zupon, who continued to manage Furolias until recently and was personally involved in the acquisition of Furolias by Erinsar, was part of the scheme outlined in our document.

We believe that the arrest of Zupon will implicate higher-ranking officers and directors of the group, and will evoke matters related in our document about Furolias, as well as other matters.

We would be grateful for acknowledging receipt of our letters and document and for advising us of the course of action that you intend to conduct."

One thing is common between the hero of this case – K or Karisios - and the hero of Kafka's 'Trial', Joseph K, - the unending optimism in the harshest conditions. When Karisios finally reached Sorial, Sorial told him that the SEC would not investigate this matter and did not even write a letter to respond officially to Karisios' complaint, although Karisios held a substantial amount of shares in a US company and was entitled at least to an official answer on his complaint.

And if Kafka, Joseph K and 'The Trial' were already mentioned, it is maybe appropriate to mention a stunning coincidence that summarizes the dilemma of the small investors and employees against the mighty 'organizations'. On the same day, April 11, 1995, two articles appeared in the Israeli newspapers. The first one was a Cantata Jubilates in praise of Gorekius, called 'A reason to be High'. 'Mr. High-Tech' as he was called in this article had every reason to be 'high'. The article told of: "The story of Memnit – a start up company that has arrived within four years to a market valuation of $800M – is a family story for Gorekius, that holds personally about 5 percent of the company's shares. The CEO of Memnit is Aran, the son-in-law of Gorekius....

(Gorekius) - 'We participated with some resources (investing together with Nalodo in Aran's company in 1991) less that $300K for 8 percent of the company... I thought that it had some merit, but I would lie if I would say by the end of 1990 that it would be such a great success. All the time I tried to hold (neutralize) myself. For all that, the VP of Nalodo, Posturck, said that it is something great and asked immediately for permission to invest $20K of his own. Finally, it was our luck that we were partners. Personally, it is indeed, a great satisfaction....'

- How did you arrive at a merger of Furolias precisely with the big sister Erinsar?

(Gorekius) – 'If not Erinsar we would have needed another partner. We arrived in 1993 to the conclusion that the losses compel us to shrink and in fact to become an R&D company only. It means, that when we would have developed new products we would have to build up all the marketing system all over again. We tried to connect Furolias to large companies all over the world... In the meantime, Erinsar looked all the time to enter the market of ...

and Furolias, with its marketing and management infrastructure, offered Erinsar exactly that. The transaction was good for everybody: for the shareholders it was good as otherwise they would have lost everything; Nalodo could cease to support Furolias as it involved a large effort for it, most of the employees kept their jobs, and it was concurrent with the strategy of Erinsar, the main subsidiary of Nalodo.'" (A reason to be High, Meuse, 11.4.95)

What are the mottos of this Cantata? The same one as in the case of the French company Loskron: we rescued our subsidiary, we kept the jobs of the employees, we were the benefactors of the minority shareholders as we did not let them lose all their investment, but only 80 percent to 90 percent of it. And indeed, in comparison to Loskron, they were benefactors, as in Loskron they lost all their investment. Not a word about the tens of millions of dollars of profits from this transaction to Nalodo and Erinsar who took over Furolias for a few million dollars, although its intrinsic value was at least $50M, as it will be proved later on.

The merger fits with Erinsar's strategy - this is 'new' compared to the statements of Gorekius and Istovius to Karisios, that Istovius was coerced by Nalodo to buy Furolias against his wish. But the most interesting revelation is that Gorekius admits that Posturck invested in Memnit $20K out of his own pocket, a statement that contradicts what Posturck told Barad that he received the money as a risk free loan prior to his due diligence on Memnit. Should this fact be proven in court, it would incriminate Gorekius, Posturck and Aran. But who would dare to sue a company valued at $800M as Memnit, and multimillionaires as Gorekius and Aran? Especially after that Karisios read another article that was published in the newspaper on the same day, 11.4.95, which was of course a sheer coincidence.

Karisios' reasons to be worried were purely subjective. He had no 'smoking gun evidence' on many rumors that he heard from friends on events related to Erinsar. Yet, the more he investigated, the more he was active on Furolias' case, some weird events happened, that he could relate only to his disclosure activities. He took his precautions, but was not deterred. He was resolved to carry on whatever the results will be. He often wondered, where has the fantastic company that he loved and admired disappeared, what has happened to his old friends that were only a few years ago ethical, shrewd, tough, yet honest negotiators?

However, he has exhausted all the modes of action and after he failed in all of them – the newspapers, the Israeli and US SEC, preliminary stages of lawsuits in Israel and the US, finding partners for an action against the organization, speaking with the auditors and lawyers of the Group, approaching the 'ruling family', trying to compromise and find a settlement, writing lawyers' letters, having friendly and menacing talks with all K's friends, and so on, he ceased his attempts as 'le combat finit faute de combattant'.

In his long route he lost many friends but he found new, trustworthy ones, who were willing to help him in need. Karisios is perceived today as a whistle-blower in some parts of the Israeli business community and it affected severely his business as a consultant. It did not prevent him from being very successful in some of his business and to earn three times more than what he had lost at Furolias. His family life was not affected, on the contrary, the Furolias trauma strengthened the family ties that were prior to that also very strong.

After analyzing the schedule of events, it is time to enter into more details on the wrongdoing of Furolias and how they wronged the rights of the minority shareholders. Once again we see that the so-called minorities were in fact the majority shareholders as they held 60 percent of the shares. Therefore they were called unaffiliated stockholders, as they were not affiliated to Nalodo, which controlled the board of directors. The best way to give the details is by presenting the 'synopsis of the statement of claim by the unaffiliated stockholders of Furolias', that was prepared subsequent to the official merger of Erinsar and Furolias on January 15, 1995. Initial versions of this synopsis were prepared during the months that preceded the merger and were showed to the holding company of Durtem, Nalodo, Erinsar and Furolias, to the Israeli SEC, to Otwuss the lawyer of Furolias, to Wasniss the second lawyer of Karisios, and to others. Those versions were meant to be sent to the SEC or to constitute a statement of claim. The final version was sent to the US SEC, to Eikon the auditor of Furolias, to Yarmuk – one of the largest newspapers in Israel, to Shiran – a governmental lawyer who tried to cooperate, and to others. The synopsis summarizes all the important facts and proves undoubtedly the wrongdoing to the shareholders, without supplying 'smoking gun evidence' for reasons mentioned earlier in the case.

"A. INTRODUCTION

On August 3, 1994, Erinsar and Furolias entered into an agreement and plan of merger after receiving the approval of their respective Boards of Directors and a fairness opinion from an investment bank concerning the acquisition, from a financial viewpoint, vis-a-vis Furolias' non-affiliated shareholders.

The merger agreement was approved by a special meeting of shareholders of Furolias International Inc. on December 30, 1994.

This claim proves that the events that preceded the merger caused severe damages to the unaffiliated shareholders of Furolias and led to a merger of Furolias with an affiliated company at an extremely low and conjunctural valuation.

All the other measures were exhausted, including talks with the management of Furolias, Nalodo and Erinsar, as well as their lawyers, but with no avail. All our questions and requests were unsatisfactory answered.

We therefore resort to this claim in order to compensate us for our damages.

B. FAIRNESS OPINION

The mandate issued by Furolias' Board of Directors to the investment banker was to give an opinion on the fairness of the transaction <u>from a financial point of view</u>. However, a financial point of view is totally irrelevant in Furolias' case, as its financial statements do not reflect adequately Furolias' potential:

Furolias has developed state-of-the-art new products with an outstanding potential, as they "integrate the newest technologies in the marketplace" according to Furolias' management. This potential is not reflected in the financial statements.

Furolias has invested in Research and Development about $30 million in the last few years which were totally expensed and are not reflected in the assets of the company.

Furolias has very valuable Know-How and Goodwill that are not reflected in the assets of the company.

Furolias' assets comprise about $20 million in Equipment, Fixtures and Improvements, but after accumulated depreciation their net assets value amounts of only about $5 million. However, their market value exceeds by far this amount. The company has one of the most sophisticated SMD equipment in Israel, which is currently used only up to 35% of its capacity.

Furolias has a customer base of thousands of satisfied customers who are an excellent potential for the new products' introduction. This, of course, is not reflected in the financial statements.

Furolias has outstanding distribution channels with three large subsidiaries in the US, UK and Germany, offices in several other key countries and distributors all over the world. The value of this organization,

which was developed at a heavy cost to the company, is not reflected in the financial statements.

\# The company has an accumulated loss of more than $40 million, that would save Furolias a very substantial amount of taxes after reaching profitability. The value of this tax benefit is not reflected in the financial statements of the company.

\# Furolias operates in a market with an annual growth rate of about 25%, one of the highest in the high-tech industry, where most of the companies are profitable, including the Israeli ones. The worldwide intelligent … market amounts to $1 billion.

A valuation based only on financial parameters would probably give Furolias a value of zero or close to it as the Shareholders' Equity is zero or negative. Therefore, commissioning an investment banker to give a valuation based only on financial parameters can lead only to results that are a priori obvious.

Only a valuation which is based on all the abovementioned parameters, as well as forecasts of profitability and cash flow of Furolias and other parameters - tangible and intangible, and is not restricted to financial parameters, although taking them into account, will give the company an objective valuation.

The valuation of Furolias would amount on the most conservative basis to more than $60 million if all the parameters are taken into account and not only the financial parameters.

The December 2, 1994 Prospectus states explicitly in page 25 that "in determining fair value, the Delaware Court is to take into account all relevant factors." What is true for the Court is true also for Furolias' investment banker, although Furolias stated in the December 30, 1994 meeting that the Board does not think it is appropriate that the investment banker will look into non-financial terms.

Knowing the composition of the Board, it is not difficult to understand why. One should bear in mind that investment bankers look in most of their fairness valuations into non-financial terms and have analysts who specialize in marketing and technology considerations.

However, we have requested to receive this investment banker's fairness opinion based on the financial parameters, but to no avail. The gist of the appraisal that was sent with the proxy is not sufficient and the excuse that the material behind it was not handed over even to the Board is cynical, as the Board mandated the banker to give an inadequate fairness opinion. The Court may ask Osttowar if as part of its activities described in page 35 of the prospectus it gives also valuations of companies that are based on non-financial considerations.

Osttowar was paid the exorbitant sum of $150,000 for rendering its opinion and Furolias has agreed to indemnify Osttowar against certain liabilities, including liabilities under Federal securities laws, relating to or arising out of services performed by Osttowar as financial advisor to Furolias...

Furolias' Board believed that these terms did not preclude Osttowar from rendering independent and objective advice. These unusual terms could be construed by the fact that it is stated in page 35 of the prospectus that Osttowar works very closely with Erinsar, that is to benefit from the merger.

The events that led to the issuance of Osttowar's opinion are described in page 31 of the prospectus. One should bear in mind that in the July 1,1994 release Furolias had committed to receive an investment banker opinion on the valuation of the company. It was expected that definitive merger agreements would be signed on or before August 1, 1994.

However, only on August 1, 1994, Furolias engaged Osttowar to give the fairness opinion. Osttowar submitted a verbal opinion to the Board only two days letter on August 3, and the written opinion on August 4. The fair valuation for the unaffiliated shareholders was therefore prepared in 2-3 days, by an investment banker who works very closely with Erinsar and has received an exorbitant amount of $150,000 and indemnification against liabilities.

Furthermore, its mandate was confined on purpose only to financial considerations, although Osttowar gives fairness opinions that are based on other parameters as well! When asked why the fairness opinion was confined only to financial considerations, Furolias replied in the December 30,1994 shareholders' meeting that "the Board does not think it is appropriate that the investment banker will look into non-financial terms!"

Osttowar is not willing probably to update its opinion and Furolias states in page 11 of the prospectus as a risk factor that Furolias does not intend to obtain an updated opinion of Osttowar at or prior to the Effective Time.

To Nalodo, the major shareholder of Furolias, it makes no difference that the valuation of Furolias in the merger transaction amounted only to about $10 million, as it has approximately the same ownership percentage in Erinsar and Furolias, but to the other shareholders it is of the utmost importance, as such a low valuation erodes more than 80% of the true valuation of Furolias.

Furolias' Board of Directors has wronged the unaffiliated shareholders by requesting a fairness opinion which was not based on the fair value of the company but only on its financial valuation. A proper valuation of the company would probably value the company by at least six times more.

C. SPECIAL COMMITTEE OF UNAFFILIATED DIRECTORS

Erinsar's offer has been reviewed and recommended by a special committee of unaffiliated directors, that consisted of Rovco and Wersnon. Rovco is an independent consultant and Wersnon is the CEO of Furolias.

The offer of Erinsar was received, studied, reviewed and approved by the special committee and Furolias' Board of Directors in nine days (June 21 to June 30). The Special Committee retained Osttowar to render an opinion on the fairness of the transaction and retained Jonroms to act as legal counsel in connection with the Erinsar Offer.

It is extremely unusual that this committee that had to protect the interests of the unaffiliated shareholders has chosen among dozens of alternatives the two firms that worked in the past very closely with Erinsar and have received from this transaction very substantial amounts. The Court could investigate why Osttowar was engaged formally to render its opinion only on August 1 and what were the activities of Osttowar with the merger prior to this date.

Mr. Istovius, the CEO of Erinsar, which is controlled by Nalodo which controls also Furolias, indicated to the committee that while Erinsar would be flexible on non-financial terms, the financial terms of the Erinsar Offer were non-negotiable. In other words, Erinsar's offer was a take-it-or-leave-it offer, a quite unusual way to treat an affiliated company, or in other terms - a friendly persuasion of a sister company with the benevolent eye of the parent company watching the transaction.

Another special committee would have probably rejected this mandate, but not our committee that capitulated after five days of negotiations and succeeded to obtain "Erinsar's responsibility to carry forward the obligations of Furolias to indemnify its officers and directors." (page 28 of the prospectus). Indeed, an excellent achievement to preserve the interests of the committee, the Board and the officers, but a complete failure in preserving our interests that the committee was supposed to preserve.

Mr. Gorekius, Chairman of the Board of Furolias, surrendered on June 29 to Erinsar's ultimatum that was expiring on June 30 and accepted to merge Furolias with Erinsar, in which he acts also as Chairman of the Board. The reason given for the hasty decision was Furolias' serious financial conditions, that Mr. Gorekius has committed to remedy throughout the whole year of 1994 as Chairman of Nalodo, that controls Erinsar and Furolias as well. Indeed, quite an awkward situation!

One should bear in mind that in early 1994, the Nalodo Board resolved to assist Furolias in meeting its projected financing needs for 1994 - (March 8, release). The interests of Mr. Gorekius, heading Nalodo, Erinsar and Furolias, converged to surrender (as Chairman of Furolias) to his own ultimatum (as Chairman of Erinsar) in order not to fulfill his obligations (as Chairman of Nalodo).

Oddly enough, we were informed in Nalodo's August 1994 prospectus that Mr. Gorekius has received from Nalodo a $800K loan on very favorable conditions to purchase Erinsar's shares, only a few days after the approval of the merger by the Boards of Erinsar and Furolias, in both of which he acts as chairman.

The Board of Directors, with a majority of affiliated directors, made probably the right decision for Nalodo by approving without delay Erinsar's offer as Nalodo's ownership share is similar in Erinsar and Furolias.

However, before making a decision that is flagrantly detrimental to the unaffiliated stockholders of Furolias, one would expect that at least the committee of unaffiliated directors would consider the offer for a period of a few weeks before taking any decisions.

One should bear in mind that the unaffiliated stockholders are the majority with over 58% of the voting power. Only one stockholder, Doherty has more than 5% of the common stock (5.6%) and the others have much less than that.

In view of that, we believe that the special committee should have requested at least a few weeks in order to consider carefully the offer, commission and receive a fairness opinion on the fair valuation of the company before making any decision.

Furolias mentioned in the December 30 meeting that Mr. Rovco is not connected in any way to the Durtem Group that controls Nalodo. We ask the Court to register this statement. Mr. Wersnon, the CEO of Furolias was appointed by its Board only in 1994, more than a year after starting to work in Furolias. He was appointed by the Board controlled by Nalodo, who gave him very generous compensation terms. Prior to then only Mr. Rovco was unaffiliated.

There is a divergence of opinions as to whether a CEO of a company controlled by a major shareholder could be impartial as well as unaffiliated, but expressly because of that he should have weighed at length Erinsar's proposal before reaching such a crucial decision in a few days. And yet, in the December 30 shareholders' meeting he is described as "really and truly a non-affiliated member."

How can Wersnon explain to the shareholders that he had voted for a valuation of $10 million for the company, while he believed all the way long that its value is much more than that.

Anyhow, Wersnon voted for ceasing the existence of Furolias as an independent entity, jeopardizing his position as CEO, and making an irreversible detrimental decision to the unaffiliated stockholders within a few days after receiving Erinsar's ultimatum.

Fortunately enough, Erinsar has committed to indemnify all the Directors of Furolias according to Erinsar's prospectus, page 75, par. 6.2.3.5 (b).

Furolias has mislead the shareholders in the December 30 meeting by stating that there was no special arrangement to protect Furolias' management against future claims and it was done according to the Delaware Law. As explained above, the management was granted full protection against lawsuits by Erinsar who is to benefit from the merger to the detriment of the unaffiliated shareholders, whose interests they were supposed to protect.

Wersnon and Rovco have wronged the unaffiliated shareholders by hastily approving Erinsar's offer which is detrimental to the unaffiliated stockholders.

They are personally liable for their hasty decision should they be found guilty for a breach of their duty of loyalty to Furolias' stockholders, for acts or omissions not in good faith or for any transaction from which the directors derived an improper personal benefit.

The DGCL does not allow Furolias in such cases to limit or eliminate the personal liability of directors to the stockholders for monetary damages for breaches of a director's fiduciary duty as a director. (page 59 of the prospectus).

Furthermore, Erinsar is not allowed to indemnify them as the Effective Time of the merger has not occurred yet. (page A-20 of the prospectus).

D. FUROLIAS' VALUATION TO ERINSAR

In the July 1 release Istovius the CEO of Erinsar is quoted as follows: "Furolias has leading technological capabilities, a worldwide distribution organization and a significant customer base in ... applications. We view their expertise and market position as strategic to our new thrust of business development in the growing ... market". This is an excellent definition for a fair valuation of the company and not a valuation based on a financial point of view.

Why should the basis for valuation of the company be different for Erinsar, owned by Nalodo by about 40%, and for the unaffiliated stockholders of Furolias?

How can Furolias' Board of Directors explain in good faith why the merger is for Erinsar an excellent business opportunity, acquiring a company valued at $60 million at $10 million, while for the others it would mean giving away their investment in Furolias at a conjunctural low price that has nothing to do with the fair value of the company.

An indication of the proper valuation of Furolias could be found in the due diligence documents prepared by Erinsar as backup for its August 15, 1994 prospectus on a rights issue, including a Furolias' chapter which is a substantial part of the prospectus.

Furolias referred to this issue in the December 30 shareholders meeting and stated that "Erinsar did not make any specific valuation of Furolias' value. If this is the case Erinsar will undoubtedly be sued by its shareholders (some of them own Erinsar and Furolias shares as well) as it contradicts explicitly Erinsar's August 15, 1994 prospectus, chapter 6.2.3.5.a, and shows negligence on the part of Erinsar to merge with a company without making any specific valuation of the company.

We request that the Court receives Erinsar's internal valuation of Furolias which was made as part of the acquisition process of Furolias and the due diligence for Erinsar's rights issue. This valuation, accompanied by a written statement of Erinsar's management that there are no other documents in this respect, could assist the valuation process for the unaffiliated stockholders.

E. PRICE FLUCTUATIONS OF FUROLIAS' COMMON STOCK

The price of Furolias' common stock suffered from an anomaly that is very difficult to explain. The price was on the average $6 in 1992 (highest- $10 7/8, lowest- $4 3/4), when the company suffered a loss of more than $25 million, and remained at an average price of $6 (highest- $9 1/2, lowest- $4 3/8), when the loss was reduced to about $6 million, after implementing a turnaround plan.

Only in 1994, the price started to drop at an accelerated rate to less than $1 for unexplained reasons, although the company has started to introduce its new products by the end of 1993, and the 1993 losses were by far less than in 1992.

In its 1992 Annual Report, Furolias' management conveyed a message of hope ending with the following sentence: "We begin a period of new growth in this challenging time of transition".

Following the dramatic improvement in Furolias' results in 1993, Wersnon stated in an article issued by Gassan on March 10, 1994, that "there is no doubt that Furolias' situation will improve in 1994. 1994 will be a good year." On this interview, inter alia, we sue Mr. Wersnon for making false statements.

Nevertheless, the company was delisted from the NASDAQ National Market on April 1, 1994. When asked on December 30, 1994, what measures did Furolias make in order to prevent the delisting, they stated that there was a meeting in Washington and the company tried to persuade the authorities not to delist, to no avail.

This is the conduct of a company that spends hundreds of thousands of dollars if not millions in order to foster a merger that is detrimental to its unaffiliated shareholders. It did not devote more than the time of a meeting in Washington. This conduct could lead to the conclusion that the delisting converged with the interests of the affiliated owners to let the price drop to less than a dollar in order to discourage the unaffiliated stockholders to hold their shares and to facilitate the merger at a very low conjunctural price.

The drop in the price was largely conjunctural, but while the price of other shares is recovering, Furolias' price remained at about $1, pursuant to Erinsar's offer, as no sensible stockholders will purchase it for more, although some curious transactions that could be investigated have occurred in the last few months with volumes of hundreds of thousands of shares.

The Court could investigate why Furolias' management did not take the necessary measures in order to prevent the delisting of the shares and if there was any negligence that was made on their part.

The result of this event was that after the sharp drop in price, the share price for the unaffiliated stockholders was set at about $1 by Erinsar's offer, leaving the prospects of the appreciation of Furolias' valuation only to Erinsar.

The Court could also investigate the conduct of the shares' price in the last few years, and who has made the transactions directly or indirectly - especially throughout 1994 as well as the transactions of key employees. Were there any interventions by the market makers of Furolias' shares and were they legal?

The Court could investigate why the July 1 release omit the fact that the shares could continue to be traded even after being removed from the Board's List of Marginal OTC Stocks, thus giving the impression that the shareholders have no other alternative but to accept Erinsar's offer, and if Furolias' management has taken all the measures in order to prevent the removal.

Furolias stated in December 30, 1994 that in the release it was clear that the trading has not stopped completely and that the stock will trade on the Bulletin Board. However, this was not stated at all in the release and the newspapers and many shareholders thought that they have no other alternative but to accept Erinsar's offer or lose the remaining value of the shares, which was exactly the interest of the affiliated shareholders.

F. FUROLIAS' TURNAROUND PLAN

Wersnon started to act as CEO of Furolias in January 1993 and has achieved a very successful turnaround plan, cutting down losses by three quarters, stabilizing the existing products, introducing successfully the state-of-the-art products to the market, streamlining and restructuring operations, laying off a large number of employees and implementing cost-cutting measures.

Furthermore, in an interview in Meuse on August 15, 1994, after the merger of Furolias with Erinsar was approved, he states that he meant to create a new Digital (DEC) and has succeeded in arousing interest in the company. How does this statement coincide with a valuation of $10 million for a company meant to become a new Digital?

In the 1993 annual report, releases and interviews, Wersnon never even alluded that he has failed in his mission and only admitted that there is a timing difference until the full impact of the introduction of the new products into the market will be achieved.

Erinsar intends to supply the bridging loans in order to complete the turnaround plan, probably a few million dollars, and will benefit from all the potential of Furolias.

Erinsar, which purchases the company for about $10 million and lends it only a small bridging loan, will have an outstanding return on investment, while the unaffiliated stockholders who have invested $6 or more per share at a valuation of above $50 million dollars will lose almost all their investment.

Furolias' management has wronged its unaffiliated stockholders by handing over the company's management to an affiliated company a short while before obtaining the results of the huge investments in R&D and marketing.

This hasty decision will give the affiliated company a remarkable return on investment, while eroding by more than 80% the investment of the unaffiliated stockholders who have backed the company and not sold their shares in the most difficult periods. Today, a few months before completing the turnaround and achieving profitability, they are deprived of the benefits of their endeavors by an affiliated party.

G. RIGHTS ISSUE

Erinsar and Nalodo have offered in August 1994 rights issues. One of the main prospects in Erinsar's rights issue is the merger with Furolias. One of the main prospects in Nalodo's rights issue is to finance the purchase of Erinsar's shares in its rights issue.

However, Furolias, which is controlled by Nalodo, has never offered a rights issue to all its stockholders. When asked at the December 30 shareholders' meeting about that, Furolias stated that the alternative was discussed with the investment banker and was rejected according to their advice.

This is the same investment banker who works very closely with Erinsar and has received an exorbitant fee for a fairness opinion, which is detrimental to the unaffiliated shareholders. Erinsar benefits from the fact that no rights issue was made in order not to allow the unaffiliated shareholders to rescue the company in which they have invested.

In this case, Furolias even saved the airline fares to Washington and did not even consult with the SEC that could have stated that this rights issue can be done, thus jeopardizing the merger with the affiliate company and giving the unaffiliated shareholders the possibility not to be diluted and preventing the loss of most of their investment.

The reason given in the prospectus, page 26, why Osttowar had advised Furolias in March 1994 that such a public rights offering was not feasible was "Furolias' poor financial performance". In the same month Wersnon states in Gassan that 1994 will be a good year.

It is impossible to understand how it was impossible to raise new equity from Furolias' shareholders at a very favorable market price, while the CEO of the company states that 1994 will be a good year. Unless, Furolias' board ruled out this possibility in order to facilitate Erinsar's approach a month later.

Furolias' management stated repeatedly that it needed an investor, while we are not in the least convinced that this was the proper solution for Furolias' problems. Furolias has decided not to raise a rights issue when the company needed only a few million dollars more as a bridging loan in mid 1994 and preferred to hand over Furolias at the extremely low price of $10 million to an affiliated party.

There is absolutely no doubt that at a price of $1 per share a rights issue would have been very successful, would have given the necessary funds to complete the turnaround plan and would have given to the unaffiliated stockholders the opportunity to retain their shares in Furolias.

The fact that Furolias' Board rejected the possibility of a rights issue is particularly curious in view of the fact that subsequently two rights issues of affiliated companies, such as Erinsar and Nalodo, were offered and Furolias appears to be one of the main assets in the prospectuses of these companies.

Furolias' management has wronged us by not making a rights issue for Furolias, thus causing Furolias' unaffiliated stockholders an irreversible loss of more than 80% of their investment. We have therefore to be compensated without any delay for all the damages caused to us by this arbitrary decision.

H. MERGER AS AN AXIOM

Furolias' full merger with Erinsar is taken as an axiom. However, we disagree completely with this axiom. Most of the turnaround plan was implemented by Furolias itself without the help of any outside parties and Furolias could implement most of the measures taken by Erinsar with its own management.

When asked why a full merger is the only alternative for us, Furolias answered on December 30 that the whole idea behind Erinsar's offer is a full merger and not a fractional one. No explanation, just a dictate!

Erinsar has signed a service agreement with Furolias at arms length conditions. However, it is not stated in Erinsar's prospectus if Erinsar will pay for the use of the 65% of the SMD unused capacity that Furolias was allowed to install at Erinsar's premises.

We have no objection that a similar agreement would prevail in the future, as long as Erinsar does not compel us to sell our shares at a detrimental price and we would still hold our Furolias' shares in the future.

In this way we could benefit from the potential of Furolias, Erinsar will purchase Nalodo's shares and other consenting stockholders' shares for 1/20 share of Erinsar to one Furolias' share and will have a majority of the company. The other unaffiliated stockholders will have the remainder of the shares.

The shares will be traded and if Erinsar manages Furolias at arms' length like it does with Soktow, none of the parties will be offended. Furolias is planned to remain a separate entity anyhow and Erinsar plans to devise an option plan for Furolias' employees, probably to purchase Furolias' shares.

Erinsar has the option to issue once again Furolias' shares that it has purchased at about $1 a share, within a year or two after having achieved profitability, at a valuation of $100 million instead of $10 million and after having recuperated the bridging loan that it has given to Furolias.

This would give Erinsar a return on its investment of 900%. The unaffiliated stockholders who have invested at a valuation of about $50 million would gain only 100% on their investment, a moderate return but still better than an erosion of 80% of their shares.

One should also add that the argument that we could benefit from the appreciation of Erinsar's shares, in case there is a full merger is completely erroneous if not cynical.

Furolias' activities will amount only to a few percents of Erinsar's activities, which are almost one billion dollars after the purchase of Deon. Furolias' shareholders do not want to be diluted in such a huge organization and do not want to receive Erinsar's shares that they could purchase at the market price without being compelled to do so.

Erinsar's proposal obliges Furolias' stockholders to sell their shares to Erinsar at about $1 per share and to buy Erinsar's shares at the current market price. They are unwilling to do both operations and prefer to remain with their Furolias' shares and to benefit directly from Furolias' potential.

It is true that Nalodo receives compensation on their Furolias' shares at a valuation of $10M, but they also benefit as shareholders of 40% of Erinsar (the same percentage as in Furolias) from the difference of $50M: between the fair value of Furolias to Erinsar - $60M and the value paid by Erinsar - $10M. Nalodo, owning 40% of Erinsar, benefits from the 40% of the $50M difference amounting to $20M, equivalent to about $6 per Furolias share on top of the $1 per share that it will receive as the other shareholders of Furolias.

This difference of $6 per share is one of the immediate damages of the unaffiliated shareholders on top of the other damages suffered by us due to the irresponsible behavior of Furolias' Board.

Furolias' management's hasty decision to accept Erinsar's proposal for a full merger has wronged the unaffiliated shareholders, by giving Erinsar 100% of Furolias' large potential for growth and a remarkable return on investment, while giving the unaffiliated stockholders no potential for growth and an average erosion of 80% of their investment.

It is proposed to leave the unconsenting stockholders with their shares or to compensate them with the full potential of their shares at a fair valuation of the company on top of the compensation on the other damages that is due to them.

I. NALODO'S FINANCING COMMITMENTS

In Furolias' March 8, 1994 release it is stated that "the Nalodo Board of Directors resolved to assist the Company in meeting its projected financing needs for 1994." This commitment was changed retroactively in Nalodo's prospectus of August 25, 1994 as follows (page 177, par. 7.2.3 b): "In view of the difficulties of Furolias, the Company (Nalodo) has decided to assist Furolias in raising the financing needed for it until December 1994."

Here, Nalodo no longer commits to assist in meeting the needs but only to assist in raising the financing needed. In this way Nalodo could argue that it has fulfilled its commitment by receiving Erinsar's offer or by reaching an agreement with the banks.

Since then Erinsar has advanced Furolias at least $3.5 million that Nalodo should have invested. Nalodo did not even propose to make a rights issue of Furolias in order to raise the financing needed until the end of the year. Nalodo hides under the formalistic statement that as it guarantees this sum it has fulfilled its commitment.

This statement is completely false as it has handed over Furolias to Erinsar on July 1 with a management agreement that left no alternative to potential investors to replace Erinsar. Should Erinsar be kept away from Furolias' management, Nalodo would have kept its promise, but as it gave Erinsar the right to manage Furolias from July 1, Nalodo could guarantee without any risk Erinsar's advance payments, as it knew that under those conditions no other investors could step in.

One should bear in mind that Furolias' management has stated that Furolias is expected to reach profitability in 1995. This will probably occur after obtaining the bridge loan by Erinsar instead of Nalodo that will enable Furolias to introduce its new products into the market.

Furolias' shareholders who have received the March release refrained from selling and/or purchased new shares, based on Nalodo's promise to meet Furolias' financing needs.

However, Nalodo has breached its promise, by transferring its commitment to Erinsar at the expense of the unaffiliated stockholders. Nalodo's attempt to change retroactively its promise would not refrain shareholders from suing Nalodo for a breach of promise that has caused them a severe loss.

J. NALODO'S ADVANCE PAYMENT FOR ADDITIONAL SHARES

According to the June 6, 1994, proxy statement, page 2 : "At March 31, 1994, Nalodo has advanced to the Company an aggregate of $2,500,000 as an advance payment for additional shares. The number of shares to be issued in exchange for this advance had not been determined as of the date of this Proxy Statement.

It is possible that, following the issuance of additional shares, Nalodo will beneficially own a majority of the outstanding shares of the Company's Common Stock."

It is quite odd that the shares were not issued on March 31, when the share's price was relatively high and Nalodo would have received a relatively small amount of shares.

One should remember that March 31 was one day before the company was delisted, which started the accelerated drop in price.

Instead of that, Nalodo chooses to wait until the date of the June proxy, when the price of the share was about $1, in order to announce that it could convert the advance payment into shares, thus receiving 2.5 million shares and obtaining a majority of the votes that it would not have obtained had it chosen to convert into shares on March 31, as it should have done.

Ultimately, Nalodo has chosen not to convert the advance payment into shares, probably because it has arrived to the conclusion that it would be sued for this flagrant usurpation of the unaffiliated shareholders' rights, although it has committed to do so, and has reached a separate agreement with Erinsar, that its legality has to be investigated.

While asked about this matter on the December 30 meeting, Nalodo states in a paroxysm of feigned naivete that it did it because it wanted to be fair to all the parties involved (including us?), with the risk of the price going up (after the delisting that it has so vehemently opposed at the unique meeting in Washington) after the plans are made public (which plans - the takeover plan by Erinsar that fixed the price to $1?). The $2.5M were lost at the end! Indeed, but they had no other choice in order to gain $20M from the merger, with their Erinsar's shares, at the detriment of the unaffiliated shareholders.

Nalodo has wronged the unaffiliated shareholders by not converting the advance payment into shares on the date of the payment and by choosing to wait until it could obtain a much larger amount of shares giving Nalodo the majority, after it knew that the price was bound to decrease due to the delisting of the shares that occurred one day after the payment.

An open issue that has to be investigated is: has Nalodo purchased since then shares or reached agreements with other shareholders, such as Doherty, and in what terms has it obtained the majority needed to approve the merger. What was the reason for Furolias' decision not to divulge the list of the shareholders, in order to prevent the organization of the majority shareholders to vote against the merger, and was it done in good faith?

K. POTENTIAL INVESTORS FOR FUROLIAS

In every occasion Furolias states that it has searched thoroughly for a potential investor but only Erinsar has made an offer. One could only compliment Erinsar for its insight that is about to bring it a remarkable return on its investment.

In the past, Erinsar has proven to possess such an insight by acquiring the majority of Soktow's shares, another affiliate of Nalodo that has incurred prior to then more than $200 million losses eroding the value of the unaffiliated shareholders from about $30 a share to about $1 a share.

In the first year after Erinsar stepped in, Soktow has reached a $13.5 million profitability and the shares jumped within a short period of time to about $6, with an unprecedented return on investment for Erinsar and a partial recovery of the losses for the unaffiliated stockholders.

In Soktow at least those shareholders, although diluted, were allowed to benefit partially from the recovery as they were not forced to sell their shares. In Soktow's case as in Furolias there were many potential investors but none had submitted a firm proposal.

It is beyond the scope of this document to find an analogy between those two cases that may prove that charity begins at home or that Erinsar is the only company with enough insight to make such excellent business decisions. The publicity of the Furolias' case may induce investors that were wronged by Nalodo who was involved in both takeovers by Erinsar to testify before the Court.

However, it should be investigated whether any company was offered to purchase all the shares of Furolias, including Nalodo's shares, at such a low price as $1 a share. Some of the potential investors could testify on the terms offered to them as well as the information and forecasts divulged to them, which may curiously prove to be substantially different from what was offered and presented to Erinsar.

One could argue that the investors could have stepped in at Erinsar's conditions after July 1, but as explained above no one was probably willing to do so after Erinsar has received the right to manage Furolias.

One should not forget that Furolias' management conveyed a confused message of continued losses with bright prospects in an uncertain future.

Only after the receipt of Erinsar's offer, the sky all of a sudden cleared and both companies answered in unison that it is an excellent deal for Furolias, Erinsar and Nalodo.

As far as the unaffiliated stockholders are concerned they should rely on the integrity of the unaffiliated directors that have made a hasty decision and commissioned a fairness opinion totally irrelevant to the fair valuation of the company, while they were not bothered by a potential class action, shelled as they are by the commitment of Erinsar to indemnify them in case of legal action against them.

Furolias' management may have handled at least some of the investors in a way that no proposal was received from unaffiliated investors. A proposal that could have been received, would have solved the financing needs of the company without resorting to a merger with Erinsar that is detrimental for the unaffiliated stockholders.

L. THE APPRAISAL INSTITUTION

Erinsar has allowed the unconsenting stockholders to apply to an Appraisal Court that would evaluate the price of Furolias' shares that will be sold at this price to Erinsar. We have decided not to resort to this institution for several reasons on top of all the reasons stated above.

The appraisal is a very costly process, especially for parties residing outside the USA, as all the costs of the process may incur to only a few shareholders. In order to present our point of view to the Appraisal Court we would need to resort to expensive lawyers and experts, thus incurring additional expenses.

The appraisal gives a fair valuation of the company, but does not take into consideration wrongdoing of the Board that caused us damages that exceed by far the difference in valuation. It is based on the merger as a fait accompli but we object in principle to a full merger.

The appraisal process could take months after the Effective Time, leaving us in a situation that we cannot benefit from our investments. One should bear in mind that as a result of the wrongdoing of Furolias' management we cannot benefit from our investment since the beginning of 1994, after the price has dropped by about 90%.

The Appraisal Court cannot give full compensation to the unaffiliated shareholders, as its scope is limited. The process is very long and expensive and would not give a solution to the wrongdoing that the unaffiliated shareholders have suffered from the company and its Board.

M. PENALTIES TO ERINSAR

Furolias' Board has committed to indemnify Erinsar on the amount of $1M in the case of termination by Furolias. This outrageous penalty is completely unjustified as Nalodo guarantees the advance payment made by Erinsar to Furolias, Erinsar is paid on a current basis on its management services and Erinsar is allowed to absorb portions of Furolias' excess manufacturing capacity, while providing Erinsar with very valuable production equipment which Erinsar does not currently possess.

When asked on the reasons for this penalty, Furolias' management answered on the December 30 meeting that the possible break-up was a great risk to Erinsar, after investing heavily in all the preparation of the merger. The penalty amount was negotiated between the companies. (probably on the same terms of the ultimatum dictate on the valuation of the company).

The outrageous penalty was probably meant to dissuade Furolias from any second thoughts on the draconian terms of the deal. Furolias could not pay this amount, as Nalodo was the only provider of funds in 1994 besides Erinsar.

Should the company have raised funds in a rights issue it could have paid the penalty, but it preferred to consummate the deal at the expense of the unaffiliated shareholders that were the only party that was wronged by the merger, as Nalodo and Erinsar benefited by a valuation of $60M to the company, the banks received collateral to their loans, the officers and directors received full protection against claims and Osttowar, Jonroms and other consultants received very generous fees.

N. FULL DISCLOSURE TO THE SHAREHOLDERS

Furolias' management kept the list of its shareholders as a secret and did not want to divulge it to the unaffiliated shareholders, fearing that they would attempt to dissuade the shareholders from voting for the outstanding deal that they have signed with Erinsar. But this obstruction is only a small misdemeanor in comparison to their wrongdoing at the June 27 shareholders meeting.

The December 2, 1994 prospectus divulges the fact that by June 27, the Special Committee had already decided to agree to the financial dictates of Erinsar and to negotiate only minor issues such as full protection from shareholders' claims in order to save the company from the catastrophe of not meeting Erinsar's ultimatum that was to expire on June 30.

Yet, not even a word was said to the shareholders who attended the meeting on June 27. Not a word on the fantastic deal for Nalodo and Erinsar that was agreed in principle, not a word to the unaffiliated shareholders on the coercion of Erinsar, not a word to give potential investors a possibility for a fair competition. An unbiased Board would have rejected Erinsar's ultimatum and enabled other investors to give a competitive offer prior to approving the deal.

This is the normal course of business instead of running amok toward Erinsar's dictate. But not our special committee. They attended the meeting, said no word to the innocent shareholders and were only preoccupied by the question - would Erinsar agree or not to compensate them in case of claims?

And on July 1 what a jubilation! We have stroked such a good deal for Erinsar and Furolias! Press releases in very optimistic tones, interviews with pictures of the two ... (Wersnon and Istovius), all the troubles are behind us, this merger will bring the company to profitability...

And what about the 58% majority shareholders, who all of a sudden have seen their investments shrink forever to less-than 80% of their investment? Who is going to protect them - Osttowar, Jonroms who work very closely with Erinsar, Wersnon who was appointed by a Board ruled by Nalodo, Rovco?

One of the main wrongdoings of Furolias was not to divulge the imminent merger with Erinsar on the June 27 shareholders' meeting. This prevented the shareholders from organizing against the merger prior to its approval by the Boards and the receipt of competitive offers. The Court will have to decide if it was done on purpose in order to facilitate the merger between the affiliated parties at the expense of the unaffiliated shareholders.

O. REQUIRED VOTE AT THE SPECIAL MEETING

The Special Meeting of Stockholders of Furolias to approve the merger was held in New York on Friday, December 30, 1994, a very convenient date if you want to obtain a minimum attendance. Pursuant to Delaware Law and the bylaws of Furolias, the affirmative vote of at least a majority of the outstanding shares of Furolias Common Stock is necessary to approve and adopt the Merger Agreement.

On September 30, 1994 there were 8,122,534 shares outstanding. As of March 25, 1994, Nalodo held 3,370,000 shares, Mr. Gorekius held 60,000 shares and Doherty held 458,400 shares. According to the December 2, 1994 prospectus no other officers and directors held any shares of the company.

This is a very unusual situation, especially in the Nalodo group where the officers and directors hold normally a few percents of the company's equity. Mr. Gorekius received his shares at the price of $0.4 per share and has a 150% profit in nominal terms from the price at the merger.

It is not clear from the prospectus if Doherty has still its shares and if Nalodo has entered into voting agreements with other shareholders and in what terms in order to obtain a majority in favor of the merger.

Furolias has managed to obtain the majority required for approving the merger. However, even according to Furolias' statement (page 23 of the prospectus) about three million shares that did not vote for the merger had the same effect of a vote against the merger. These shares constitute about two thirds of the unaffiliated shareholders that voted in effect against the merger.

In similar cases a vote of at least a majority of the unaffiliated shareholders is required in order to approve a merger between affiliated parties. But, here again Furolias is an exception. Should the unaffiliated shareholders who had voted for the merger known all the facts that are presented in this claim, they would have voted against the merger, unless Nalodo or another related party would have made special agreements with them.

As Nalodo has almost a majority of the shares, it did not think suitable to request a majority of the votes of the unaffiliated shareholders. If it would have done so, it would have failed to approve the merger by the meeting.

In page 30 of the prospectus, Furolias deals with this issue and states: "The Furolias' Board considered requiring a vote of unaffiliated stockholders to approve the merger but concluded that such a vote would be inappropriate in light of Furolias' serious financial condition and the importance of the merger to Furolias' continued viability."

Nalodo, controlling the Board, has nominated itself as the patron of Furolias' financial condition and viability. This presumptuous attitude of a minority shareholder implies that the unaffiliated shareholders do not care for Furolias' financial condition and viability. We were not offered a rights issue because Osttowar, that works very closely with Erinsar, did not recommend it.

We were not offered to influence the voting because Nalodo, that was to benefit from the voting, thought it was inappropriate. Probably, Nalodo would state that it acted in good faith when the members of the Board of Furolias disregarded the fact that Nalodo vouched for the financial condition of Furolias and was to gain from the fact that its commitment was to be transferred to Erinsar, where it holds 40% of the shares, as in Furolias.

Furolias' Board decision not to require a majority vote of the unaffiliated shareholders that is common in such cases, as we are the majority of the shareholders, enabled the meeting to adopt the resolution in favor of the merger that would have been otherwise rejected, as the results of the vote shows. This decision favored flagrantly Nalodo at the detriment of the other shareholders and caused us very severe damages.

P. CLAIMS AND DAMAGES

We are holding respectively 111,700 shares, 48,000 shares and 7,000 shares of Furolias and claim the damages inflicted to us by Furolias, its Board of Directors, its Special Committee, its Officers and Nalodo the amount of $40,000,000. We sue the companies and the individuals personally for the whole amount and request the Court to award us the amount of $40,000,000 as damages for all the wrongdoing committed to us.

Q. CONCLUSION

All the evidence presented in this claim brings to the conclusion that due to a sequence of wrongdoing by Furolias, its Board, Special Committee, Officers and Nalodo, the unaffiliated stockholders were offered a deal which is obviously to their detriment, while being very profitable to affiliated parties such as Nalodo and Erinsar.

We have proven our case without resorting to information that was not made public. We have not dealt with a potential premeditation by Nalodo of a takeover of Furolias, we have not elaborated on a potential pattern of action with analogies to the takeover of Soktow by Erinsar, we have not given statements of potential investors who were discriminated against in comparison to Erinsar.

We have not dealt with issues of misrepresentation of financial results of Furolias by its officers, we have not analyzed the cause of Furolias' heavy losses, we have not given indications on the reasons of the shares' price collapse, we did not elaborate on the utilization of insider information in order to make profits and prevent losses in shares transactions, and other issues.

We reserve the right to do so at a later stage.

This claim is against very influential bodies that stamp arrogantly on the unaffiliated stockholders' rights. These bodies, backed by the best lawyers and consultants, try to legitimize wrongdoing by a very sophisticated mode of action. However, we are confident that the Court will succeed to unveil the mask of hypocrisy of these bodies and to render justice to the offended parties."

A lot of eloquence and no results! The proof is that although this document, which shows in flagrant terms the wrongdoing to the unaffiliated stockholders, was sent to almost every one concerned none of them thought that it was worth any consideration and every one found an excuse to back off. The main problem of this document is that the recipients had no interest in handling the matter. Even the journalist, who probably did not understand the intricacies of the affair, told Karisios: 'Bring me a juicy story about a Yeshiva director who rapes his pupils, this will sell the newspaper much more than your story!' And he sent to Erinsar a list of questions about the transaction that joined the existing lists that Karisios' lawyer Yraye sent to Furolias, and that Umberto asked at the December 30 meeting. So many questions and so few answers...

By the end of 1996, after exhausting all the possible moves, Karisios could still hear the 'innocent' answers of Wersnon, when he phoned him desperately at the eve of the announcement of the merger after he learned of the forthcoming events. Wersnon, his friend and employer, who was a member of the Special Committee of the unaffiliated directors and was supposed to protect him as a man of integrity, as a friend: "Nothing was disclosed (on the June 27 shareholders' meeting, when all the mergers details were already agreed upon effective as of July 1). Let me think (pause), nothing special, the standard stuff, directors and matters, Lortow auditor, things like that. We talked about how we are searching strategic alliances and working hard on that... In the last six months, I am working on that like crazy. If not 15 or 17 different companies, we did a tremendous job, and I think that we have achieved something... there is something that has to be published and you will be able to know... I am trembling terribly..."

After his long Odyssey, Ulysses Karisios rested his case!

Three years after Karisios' letter no judgment was issued on Zupon's case. Erinsar benevolently hired Zupon a lawyer, thus ensuring probably his cooperation. There is no chance that Zupon, who was the CEO of Furolias after its acquisition by Erinsar, will ever speak, if he has something to say at all, and the truth about Furolias will probably never be known.

Justice was not done, but the Erinsar case to follow undermined even more the image of the management of Erinsar and Nalodo.

Some of the operations, assets and liabilities of Furolias were sold by Erinsar in September 1996 to the US company Mastoss for $22.8M, more than twice the acquisition price for the whole company, as will be described in the Mastoss case. This shows that the too humane conduct did not give a return on investment of 900 percent, as hoped, buy only a ROI of at least about 150 percent in less than two years! If they would have learned from Loskron, they would have not paid at all for the company and achieve an infinite ROI. But Gorekius and Istovius' conduct was never completely unethical, as they tried always to keep up appearances.

Gorekius has retired, 10 years after the official retirement age and with tens of millions of dollars in his banking account. Istovius lives a substantial part of his time in the US and has resigned from Erinsar. After his exploits in Furolias, that won his masters tens of millions of dollars, and in Erinsar, as described in the following case, that won them hundreds of millions, Israel is probably too small for him. Could he be planning in the US exploits in the billions? Nevertheless, he does not have to be afraid of the future with a bank account of tens of millions dollars.

To conclude, if the SEC is not the right answer and the Israeli newspapers were not interested in the case that was too complicated to understand, what could we do legally in order to find an equitable solution? Apparently, it is very easy to crumble all this affair. It suffices that one witness, only one witness, (maybe Zupon), will testify against them, but all the parties are so involved in the organization that until now none of them is ready to give testimony. They are lured by the high wages, they are afraid of the consequences, they are unwilling to mess up. Only in a different climate things might change, only in an ethical environment people will not be willing to suffer any more from those evils. The conclusion is that probably this book and similar ones, ethical codes, newspapers editorials, the Internet, activist associations like ADAM, with a specific mandate to safeguard the rights of the minority shareholders, but most of all transparency would be able to change in the long run the present climate in favor of those shareholders and attenuate the wrongdoing of the corporations toward them.

10
CASE STUDY OF THE ISRAELI COMPANIES ERINSAR AND SOKTOW

"Hippolyte – Quelques crimes toujours precedent les grands crimes.
Quiconque a pu franchir les bornes legitimes
Peut violer enfin les droits les plus sacres;
Ainsi que la vertu, le crime a ses degres;
Et jamais on n'a vu la timide innocence
Passer subitement a l'extreme licence."
(Racine, Phedre, Acte IV, Scene II, 1094-1098)

"Hippolyte – Some crimes always precede major crimes.
Whoever has crossed the legitimate borders
Can ultimately violate the most sacred rights;
As with virtue, crime has its degrees;
And never have we seen timid innocence
Cross over suddenly to extreme license."

This case develops like a self-fulfilling prophecy, after the case of Furolias, where the writing was on the wall and Karisios predicted to the SEC and everybody else what will be the outcome of the silence of the lambs on Furolias. But before entering into the intricacies of the case, one should analyze the excellent article of Amar Bhide and Howard H. Stevenson "Why Be Honest if Honesty Doesn't Pay": "Even those with limited power can live down a poor record of trustworthiness. Cognitive inertia – the tendency to search for data that confirm one's beliefs and to avoid facts that might refute them – is one reason why. They don't want references or other reality checks that would disturb the dreams they have built on sand... Even with a fully disclosed public record of bad faith, hard-nosed businesspeople will still try to find reasons to trust...

Aggrieved parties may underplay or hide past unpleasantness out of embarrassment or fear of lawsuits. Or they may exaggerate others' villainies and their own blamelessness. So unless the victims themselves can be trusted to be utterly honest and objective, judgments based on their experiences become unreliable and the accuracy of the alleged transgressor's reputation unknowable. A final factor protecting the treacherous from their reputations is that it usually pays to take people at face value. Businesspeople learn over time that 'innocent until proven guilty' is a good working rule and that it is really not worth getting hung up about other people's pasts. Assuming that others are trustworthy, at least in their initial intentions, is a sensible policy...

Today's model citizen may be yesterday's sharp trader or robber baron. Trust breakers are not only unhindered by bad reputations, they are also usually spared retaliation by parties they injure. Many of the same factors apply. Power, for example: attacking a more powerful transgressor is considered foolhardy... Getting even can be expensive; even thinking about broken trusts can be debilitating. 'Forget and move on' seems to be the motto of the business world... The loss suffered through any individual breach of trust is therefore relatively small, and revenge is regarded as a distraction from other, more promising activities. Retaliation is a luxury you can't afford, respondents told us. 'You can't get obsessed with getting even. It will take away from everything else. You will take it out on the kids at home, and you will take it out on your wife. You will do lousy business.' 'It's a realization that comes with age: retaliation is a double loss. First you lose your money, now you're losing time.'...

Without convincing proof of one-sided fault, the retaliator may get a reputation for vindictiveness and scare even honorable men and women away from establishing close relationships. Even the cathartic satisfaction of getting even seems limited. Avenging lost honor is passe, at least in business dealings. Unlike Shakespeare's Venetian merchant, the modern businessperson isn't interested in exacting revenge for its own sake and, in fact, considers thirsting for retribution unprofessional and irresponsible... Assessing the value of protection against the loss of power is even more incalculable. It is almost as difficult to anticipate the nature of divine retribution as it is to assess the possibility that at some unknown time in the future your fortunes may turn, whereupon others may seek to cause you some unspecified harm. With all these unknowns and unknowables, surely the murky future costs don't stand a chance against the certain and immediate financial benefits from breaking an inconvenient promise. The net present values, at any reasonable discount rate, must work against honoring obligations...

Our tolerance for broken promises encourages risk taking. Absent the fear of debtors' prison and the stigma of bankruptcy, entrepreneurs readily borrow the funds they need to grow... We 'adjust' – and allow great talent to offset moral frailty – because we know deep down that knaves and blackguards have contributed much to our progress. And this, perhaps unprincipled, tolerance facilitates a dynamic entrepreneurial economy." (Bhide Amar and Stevenson Howard H., Why Be Honest if Honesty Doesn't Pay, Ethics at Work, Harvard Business Review, p. 48-53)

This case reflects exactly the state of mind of the businessmen as described in the article. They expected after winning the case of Furolias that everything is permissible and that people will not respond to them vindictively, as it is not businesslike. You incur your losses and continue further on to the next opportunity or to the next scam. After all, suckers don't die, they are just replaced by new ones... However, as with Karisios in the case of Furolias, they encountered a troublemaker, Astossg, the CEO of Soktow, the main subsidiary of Erinsar. Astossg as a CEO had much more public exposure and conducted a crusade against Gorekius and Istovius. He even sued them in a class action for $100M and disclosed all the wrongdoing of Erinsar and Nalodo, and their CEOs Istovius and Gorekius. This campaign did not prevent them from concluding the dealings that brought to their companies and to them personally huge rewards at the expense of the 'expendable' minority shareholders.

The article and similar ones legitimize the conduct of the businessmen like Gorekius and Istovius, who are perceived as entrepreneurs with impeccable reputation, who were never convicted, even if rumors about their lack of ethics are getting stronger and stronger. Only ethic 'fanatics' like Astossg succeeded in tarnishing their reputation but at a tremendous personal cost. But even if Gorekius and Istovius have stepped out of the scene, their masters, the families owning the large organizations and ripping most of the profits, will never step out. The Gorekiuses and the Istoviuses are expendable, at a high cost of course, with parachutes of millions of dollars, but those who control their companies and tell them what to do and remain far from the public eye are never sanctioned, as the buck stops at the Gs (Gorekiuses) and Is (Istoviuses), hereinafter the GIs.

In order to understand the environment in which the Israeli companies operate, it is necessary to analyze the centralization of the economy in Israel. In fact, most of the Israeli economy, at least the companies that are traded at the Israeli Stock Exchange, is in the hands of about 20 rich families, who practically 'own' the economy. Hundreds of articles have been written against this centralization which gathered its momentum in the last ten years after the privatization of the Government's and Labor Unions' owned companies. According to the Israeli newspaper Yarmuk of August 2, 1999, The Durtem group, the parent company of Nalodo, Furolias, Erinsar and Soktow companies mentioned in our cases, owns 22.4 percent of the Israeli Stock Exchange, with 45 companies controlled fully or partially and a market valuation of more than $10 billion. This group is controlled practically by one family. The second family controls 7.4 percent of the market, 14 companies with a market valuation of more that $3 billion. The third family controls 6 percent of the market with 12 companies and slightly less than $3 billion. The fourth – 3 percent, fifth – 2 percent, sixth – 2 percent, in the seventh place we have an international organization with 2 percent, and then in the next 20 places with 1 percent and less, practically all companies are owned by wealthy families. Israel may be a high-tech superpower but as far as capitalistic centralization is concerned, more and more people call it a 'banana republic'. This power held by a few people has tremendous implications for the ethical conduct toward the minority shareholders of those companies, who in most cases do not belong to the ruling families. Furthermore, when courageous individuals like Karisios in the Furolias case and Astossg and Poftrim in this case do dare to fight against the Durtem Group, they are confronting the superpower which controls one quarter of the Israeli stock exchange and has very close business contacts with the other 20 ruling families.

The reader of this book may be baffled by the intricacy of the plot, and the huge amount of details given in the cases of this book. It is sometimes difficult to follow and understand the meaning of the events, especially to people who do not come from the financial milieus and who have no experience in such cases. Nevertheless, this is the way that wrongdoers in the business world act, they try to complicate so much their schemes so that nobody will understand them, the public, the SEC, the auditors, or the press. In this way the minority shareholders are lost in the multitude of the details and cannot fight back. The cases are, therefore, the summary and a simplified version of tens of thousands of pages, in thousand of documents, financial reports, balance sheets, Internet posts, press releases, articles, and so on that were read and analyzed in order to prepare those cases.

The Erinsar and Soktow case is probably the most intricate, but the version given in this case is simplified, as the detailed events are much more complicated and only very sophisticated investors, or long-time insiders, would understand them and perceive the hidden means of those events. Yet, probably no book has treated in such detail and profound analysis the wrongdoing committed to minority shareholders. Maybe no ethicists have encountered such problems, maybe they are not interested in the subject in spite of its relevance to millions of shareholders, maybe they are too academic and do not analyze in depth real life cases as in this book's cases.

It is the privilege of the author of this book to have worked for more than 30 years in this business environment, knowing all its ramifications, and he asks for the forbearance of the readers if the real life vividness of the descriptions comes sometime at the expense of the academic purity. Unfortunately, real life business is not so pure and precise as academic work, and these entrepreneurial cases as well as this book should be treated leniently and with the adequate patience given to pioneer research works. The benefits that might ensue from the publication of this book might outweigh by far its defaults, as it might lessen the wrongdoing made to minority shareholders, who were not spoiled too much by academic attention until now. And after all, the minority shareholders are all of us, especially since the pension funds invest so heavily in the stock exchange.

In 1994 Erinsar purchased Deon, an international company in k... high-tech, selling about $200M annually, for the very low amount of $70M due to Deon's losses. This purchase followed Erinsar's purchase in 1990 of an ailing subsidiary of Nalodo, Soktow, in the same product line of Deon. Erinsar was prior to then quite successful in its product lines which were mainly in s... high-tech with some small r... high-tech operations. Soktow was until 1985 very profitable, was traded at the New York Stock Exchange, and had about the same sales turnover as Erinsar. All of a sudden, in 1985, the management of Soktow disclosed to the bewildered shareholders that it had incurred losses of above a hundred million dollars. The shares' price that was traded in the past at about $30 fell immediately very sharply and reached ultimately a price of less than $1. The Israeli government assisted the company in the past and following the losses of Soktow in a package of benefits worth more than $100M.

Most of the loans of the Israeli Banks, most of them shareholders of Nalodo, to Soktow were foregone, Nalodo came to the rescue of Soktow in its immediate cash needs, Suram, the CEO of Soktow resigned and left for the US as he was afraid of being arrested in Israel for fraudulent acts of withholding information from the shareholders. Gorekius, who was a member of the board of directors of Soktow, managed not to be blamed for what happened and the CEO took effectively all the blame. In retrospective, it could seem improbable to conceal such huge losses that were accumulated over a large period of time from your parent company and board of directors, although it is quite easy to conceal them from your shareholders.

By 1990, most of the turnaround of Soktow was accomplished, and the company faced a renewed profitability. This was probably known to Nalodo, the parent company of Soktow and its CEO - Gorekius, to Erinsar the sister company of Soktow and its CEO – Istovius and to the insiders, including Soktow's CEO – Poftrim. The Israeli banks held a large amount of shares of Soktow, received in lieu of the foregone loans, but were not effective in controlling the company and were interested in selling their shares even at the then current market price of less that $1. Many investors were interested in purchasing Soktow at the attractive valuation of tens of millions of dollars but they did not know exactly what was the status of the recovery of the company. The executive in charge of handling the investors was as usual Gorekius, who managed also in this case, as in the case of Furolias, to arrive to the outcome that none of them submitted a formal proposal. A friend of Karisios, Doherty, assured him that he had found a large company that was interested in purchasing Soktow but Gorekius did everything to discourage the investor and that ultimately the investor sent a complaint letter to the company.

This 'patent', which Karisios called the Gorekius patent, inducing all the investors interested in Soktow and in Furolias to lose courage after being conditioned by Gorekius and selling the subsidiaries of Nalodo to another subsidiary, Erinsar, is a very sophisticated patent that brings huge capital gains both to Nalodo and Erinsar. This patent was in fact invented by Gorekius in the early '80s when Candor, the US owner of 50 percent of Erinsar's, sold effectively its shares to Nalodo in 1981 based on the low valuation of Erinsar that incurred heavy losses in 1980. Immediately afterwards came the turnaround of Erinsar, that Istovius and Karisios managed, which increased Erinsar's shares price from about $0.5 to $13, most of the benefit going to Nalodo at the expense of Candor which left just before the turnaround. This patent is very simple: You take advantage of the losses and the low valuation of your subsidiary just before the turnaround that you know is coming. You convey a message that the situation of the company is desperate, and you do your utmost to discourage potential investors, partners and minority shareholders. If shareholders sell their shares all the better as the price goes even lower and you benefit more when you make the turnaround and the price of the shares and the valuation of the company increases.

You count on the fact that your partners are ignorant of the insider information that you have, whether your partners are like Candor who are far away in the US, or the Israeli banks, which do not know how to control effectively an ailing company and that are content from the fact that anyhow they are going to benefit from the turnaround as shareholders of Nalodo, or the potential partners who do not know of the coming recovery and are conditioned by you that the situation is beyond despair and anyhow you do not intend to sell Nalodo's shares in the company, so that they will be incapable of gaining full control of the company and implementing freely their policy.

Karisios noticed that the Gorekius patent succeeded beyond expectations in the takeover of Erinsar from Candor by Nalodo, in the takeover of Soktow by Erinsar, and in the takeover of Furolias by Erinsar. In fact, Karisios has figured that most of the profits of Nalodo throughout the years did not come from technological inventions of the company or its subsidiaries but from such exploits that came at the expense of the minority shareholders and the partners. Quite a long way from the high-tech dream of Gorekius who became a shrewd financier and transferred most of his management attention to the US stock exchange, spending a large portion of his time in New York and knowing most of the investment bankers personally. One could ask, 'but the investment bankers have lost also from the fluctuations of the price of the shares of Nalodo's subsidiaries'. A careful examination of the facts show that in fact it was their clients that incurred the losses (a big difference in the Wall Street's code of ethics), assuming that the bankers did not possess themselves insider information supplied by Gorekius or an 'insight' as had the fund of the Durtem Group, owner of Nalodo, which sold its shares of Furolias exactly before their price fell sharply from $6 to $1. The investment bankers had very high profits from their fees as the underwriters of the many successful public offerings of Nalodo and its subsidiaries, from their consulting to the companies as we learned in the Furolias case, and so on, and if some minority shareholders who sold subsequently their shares lost most of their investment, it was attributed to the risks of investment in the high-tech industry.

Erinsar purchased the shares of Nalodo and the banks and became the major shareholder of Soktow in January 1990 with more than 60 percent of the shares. As the shares were bought for less than $1 a share, Erinsar effectively paid 32.1 million dollars for a company selling about two hundred million dollars. The company made a miraculous turnaround and became profitable immediately after Erinsar took over, maybe because of the sound economic guidance of Istovius who was appointed as Chairman of Soktow, receiving a very generous amount of shares - 700,000 - at less than a dollar a share for a consideration of $469,000 received as a loan. This could have enabled him to earn about $3M when selling at least 450,000 of the shares in 1993 (he remained with 241,340 prior to 31.7.93), at the end of the restriction period, at their peak of about $8 a share. In effect miraculously the shares' price started to drop after the end of the restriction period of the shares in 1993 and they shrunk by about 75 percent within a year. This brings us to what Karisios called - Istovius' law, which is – if you are a minority shareholder and want to win risk free windfalls – buy your shares when Istovius or the GIs receive their shares or warrants and sell them when they are allowed to sell and effectively sell them. You are almost certain to make more than 1,000 percent return on investment risk free. It is a sheer miracle how in most of the cases, Erinsar's shares were at their bottom when the executives received the shares or warrants and reached their peak when they were allowed to sell them, dropping immediately afterwards very sharply, and so on. Istovius' example with Soktow's shares is one of the many examples.

In 1994 Soktow's profitability declined and Erinsar purchased Deon with a similar sales turnover as Soktow in order to increase the Group's critical mass in the k. high-tech. Throughout the years Soktow remained an independent entity, as Erinsar did not want to invest the high price of purchasing the remaining 40 percent shares that were held by minority shareholders. On the contrary, Erinsar's ownership was diluted to about 55 percent. You cannot make tremendous personal gains from the increase in valuation of Soktow and in parallel take it over at a low price. You can't win them all. But one should not despair, as this case will prove. Poftrim, the CEO of Soktow and a friend of Karisios, did a fantastic job in recovering the company but limited the involvement of Istovius into the company to Istovius' role as Chairman of the Board. Poftrim ensured that Erinsar did not have any role in Soktow and it behaved only as a shareholder. However, the relations between the two executives, Istovius and Poftrim, deteriorated over the years and after the decline in profitability of Soktow and the decision of Istovius not to amalgamate Deon's and Soktow's operations, Poftrim decided to resign and started a campaign against Istovius and Erinsar in the newspapers. He accused them of unethical and unlawful acts against the shareholders of Soktow and especially that Istovius did not fulfill his promise to merge Deon and Soktow.

Istovius had probably no intention of doing so, even if he promised that, as Erinsar had 100 percent of Deon's shares which he bought at a very low price and only 55 percent of Soktow's shares. Istovius was confident probably that Deon's profits would increase sharply and he did not want to share this profitability with the minority shareholders of Soktow. The official excuse was that Deon's management was not willing to work with Soktow, quite an awkward excuse after stating that the purchase of Deon was meant to streamline the k. high-tech operations and to achieve economies of scale. Besides, Soktow did not have the funds to purchase Deon, but Erinsar, although it had the funds preferred to finance most of the acquisition through a rights issue in 1994, the same one that was meant to finance the acquisition of Furolias. Why Istovius did not make a rights issue of Soktow to finance the purchase of Deon and preferred to make a rights issue of Erinsar is obvious in view of the considerations stated above. Poftrim was also a shareholder of Furolias, as Karisios, and did not sell most of his shares in Soktow as Istovius did in 1993, as he believed in the future of the company and wanted to give an example to his employees. The campaign of Poftrim proved unsuccessful and nobody was bothered by the arguments he raised. He warned the shareholders of the future of Soktow under Istovius' direct control but to no avail.

In 1996 Erinsar decided to split its shares into three companies – the s. high-tech company (the original Erinsar), the k. high-tech company (comprising mainly of 55 percent of the shares of Soktow and of Deon) and the r. high-tech company (comprising mainly of Furolias and a Toren division). Erinsar became effectively Nalodo, as Nalodo did not have many other important activities, besides the shares of Memnit, some other low percentage ownership in companies, and some start-ups. But Nalodo had still to play a decisive role in the future events, benefiting from the exceptional gifts of Gorekius as a financier. Karisios warned the US SEC of the outcome of Erinsar's policy, as Poftrim warned the public in the newspapers, but all the warnings did not prevent Gorekius and Istovius to implement what Karisios thought was their long range planning. In 1998 Erinsar sold Deon to Gosstik at a price of $228M. Karisios knew that Erinsar did not succeed in effectively turnaround Deon, in spite of the tremendous efforts of Istovius. The success of Istovius in selling this failing company at a price three times higher than what Erinsar bought it four years ago was astounding. Especially in view of the fact that Gosstik is one of the largest corporations in the world selling tens of billions of dollars.

Incidentally, at a wedding of the son of a mutual friend, Karisios congratulated Istovius for his tremendous achievement. Karisios asked to be seated among his new friends as he did not want any contact with his good old buddies. It did not prevent them from attracting him to their table, where he was surrounded once again with his old friends. He looked around him and saw Istovius, Hustash, Bsosskins, seated next to him, and Gorekius standing above him. It was the first time in four years that he saw most of them. He looked at them and said: 'Now I feel secure!' A very pleasant conversation ensued, after all friendship is thicker than blood. Istovius was very cordial, asked him what he was doing and when he heard of his activities in ethics he told him: 'If you want to interview an unethical CEO come to me' with a sense of humor that delighted all the participants in the conversation. Karisios praised Istovius for the sale of Deon, and said: 'This time you succeeded in manipulating a large corporation and not just miserable small shareholders. Congratulations! You have climbed to the highest league.' Everybody laughed but Karisios could not know at this moment that they were laughing of his innocence as he learned a few weeks later.

Istovius replaced in 1994 Poftrim by Astossg, the former VP Sales of Soktow. Astossg, a friend of Karisios and Poftrim, was suspected by them of cooperating with Istovius in the attempt to wrong the minority shareholders of Soktow. Poftrim cut abruptly his contacts with Astossg, Karisios remained cordial with Astossg but could not believe that an ethical man could cooperate with Istovius at this stage. Astossg told Karisios after the events that will follow in 1998 that he believed that Istovius really meant to fulfill his promises about Soktow. There is probably no limit to what a man is willing to believe, especially when his counterpart is such a charming and convincing man as Istovius, and you are offered the position of CEO of the company in which you worked for all your professional life. After all, Karisios was fooled himself by Istovius and Gorekius several times and he was the last one to throw stones at others. We have to understand that every time Gorekius and Istovius raised a level in their unethical conduct. Karisios did not see any fault in Gorekius' conduct toward Candor when Nalodo purchased their shares in Erinsar. He presumed that Candor, a multibillion US company, would be smart enough to know what was happening. Yet there were other executives who left Erinsar and Nalodo, prior to Karisios' resignation as they were more sensitive to their wrongdoing. He was also probably perceived by those people as unethical until the deeds of Istovius forced him to leave as he could not cope any more with the increase in his unethical conduct.

In the same manner, Poftrim, who was an honest man, cooperated with Istovius for four years until he too could not suffer any more Istovius' unethical conduct and he left. The same sequence of events happened also to Astossg, who was honest too. He believed that what Poftrim told him were exaggerations or he preferred to believe so, until the events showed him that Poftrim was right and Astossg also was compelled to leave. A friend of Karisios once told him a Romanian proverb: 'The shirt is close to the body, but the skin is even closer'. When you are far from the events you do not sense their full implications but when you are in the middle of the storm you are bound to notice what happens as this time your skin is at stake and not just your shirt. There are no completely ethical or unethical men or women, it is just a question of degree. Yet these cases and this book show that for most people there is a minimal degree of ethics that they cannot forego if they do not want to betray their integrity. Only executives who remained with Istovius and Gorekius throughout most of their career, knew what was happening and did not react, could be perceived by those who left the organization as almost completely unethical. But fortunately, many of the top executives of Nalodo, Soktow and Erinsar left at one stage or another.

Erinsar sold Deon to Gosstik on February 13, 1998. For simplification purposes this case will continue to refer to the k. high-tech company after the split as Erinsar, although Erinsar split into three different companies. But as this case deals only with the k. Erinsar it will be referred to as Erinsar. Unofficially, Istovius continued to pull the strings of all three companies, so that Erinsar continued to exist and the new Es continued to be identified with Istovius, or to be more specific with the hands of Istovius and the spirit of Gorekius, the CEO of Nalodo, the parent company of all the Es. The split was a brilliant financial move as it increased the valuation of Erinsar by more than twice without a substantial change in the financial results. Furthermore, it enabled Istovius and Gorekius to perform the ultimate exploit, the sale of Erinsar (k.) as will be explained later on.

On March 6, 1998, Erinsar (k.) published the results of 1997 – Revenues of $493M in comparison to $525M in 1996 and Net Income of $4.8M compared to $8.1M. This decline in revenues and sharp decrease of 40 percent in profitability explains why Istovius sold Deon. The purchase of Deon was not successful operationally, which shows that Istovius and Erinsar were very weak in operations but were excellent in financing and negotiations. Soktow's results were also disappointing – a decrease of revenues from $311M in 1996 to $303M in 1997 and in Net Income from $8M in 1996 to $0.7M in 1997. Soktow was practically losing money while Deon earned about two percent of revenues, assuming the other k. operations were marginal. No one can tell what would have happened if Istovius would have merged the operations of Deon and Soktow as promised to Poftrim and Astossg, but it did not matter to

Erinsar and Gorekius, who managed to take advantage of these setbacks as they always did, however at the expense of the minority shareholders of Erinsar and Soktow.

Erinsar (k.) traded by then at about $7 a share. Erinsar had in cash $230M less the taxes and the transaction costs. Erinsar announced on April 17, 1998 that the net gain will be only $103M (a totally incomprehensible low amount, due to allegedly high provisions for taxes and costs, which proved $50M higher in the right timing), and Nalodo's share of it with about 40 percent of the shares of Erinsar was about $40M. Still, Erinsar's share price was very attractive, but the share's price did not increase accordingly. The financial community, accustomed to Erinsar's exploits probably gave a discount as they knew that they will not benefit from the cash reserves as the insiders. They proved to be right. By 1998, the Internet became the vehicle for spreading information, press releases, stock talks, rumors and misinformation. The minority shareholders had at last a vehicle for learning a lot more about the company. It was not perfect, but it was far better than what they had before. Still, the attendance to the stock talks of Erinsar and Soktow, both of them traded in the US, was far smaller than the attendance to a smaller company as Mastoss, that will be treated in the next case. No smashing discoveries were made about the companies, and the relevant information that was shared was minimal.

In March 1998 the valuation of Erinsar was $150M,with 21,450,000 shares at a price of $7. The price fluctuated in 52 weeks between $8.5 to $4.5. The net equity was $181M, so the valuation was close to the equity and the Price to Earnings ratio was about 30. Soktow's share price was in the same period $7.5. No, it did not increase by 700 percent, the company made only a reverse split of 5 to 1, and the equivalent price was therefore $1.5. Between the splits and the reverse splits the shareholders were completely confused, which was probably what Erinsar wanted to achieve. The price varied between $9 to $6.375. The valuation was $119M and the net equity was $199M. Here, the market valuation was 40 percent lower than the equity and the company traded at a very large discount.

The board of directors of Erinsar was unusual as it was comprised of four affiliated shareholders – Istovius, Chairman, President and CEO, Gorekius – CEO of Nalodo the parent company with 40 percent of the shares, and two members of the parent company of Nalodo. Four unaffiliated directors were on the board, very respectable and honorable Professors with impeccable reputation. The unaffiliated shareholders of Erinsar could sleep in peace as nothing could happen to them with a balanced board with such respectable figures. But after knowing well the organization, one tends to believe that this was exactly the intent of Gorekius, to have a balanced board that nobody will ever dare to question its decisions should he decide to carry on his exploit, as will be explained later in the case. One could only admire the sophistication of Gorekius in manipulating the board and achieving the best results for Nalodo, even if they are to the flagrant detriment of the other shareholders, and to hide behind the respectability of the board, exactly as he hid under the culpability of the CEO of Soktow in 1985 without tarnishing his reputation. The auditors of the company were still our acquaintances of Furolias, Eikon – Karisios' old friend but not anymore, who was a senior partner in Lortow – one of the largest auditors' firm in Israel, a member firm of Ascorage, one of the big five.

On September 1, 1998, 'nu techie' writes in the Yahoo! Finance thread of Soktow - 'Wow, they sacked the CEO. Trouble ahead.' Astossg, the CEO of Soktow, opposed the sale of practically all of Soktow's activities to Priam and Gosstik. The sale was finalized on November 27, 1998 and the company announced the sale of its … business to Priam for $269.5M and the sale of its … businesses to Gosstik for $100M. Based on its management's estimations, the net proceeds from the two transactions were expected to be between $200M and $240, and net earnings for Soktow of at least $50M. Erinsar (k.), the parent company of Soktow was expected to record a gain of at least $28M, and Nalodo, the parent company of Erinsar, would record a gain of no less than $11M. One should pay notice to the wording of the announcement. We have learned from the Furolias case that the wording is expected to leave a certain impression on the shareholders, while remaining within the borders of the law. When it is said that the net proceeds are expected to be 200-240, we should understand that it will be $240M, as the taxes and transaction costs would exceed $130M (270 + 100 – 240) only in a very pessimistic scenario. When it is said that earnings will be at least so and so, we should understand that in fact it will be much more but they want to leave the impression of limited profits. The official explanation of this extremely cautious announcement is the traditional conservative approach of the Group, but the practical explanation is probably that they do not want the shares' price of Soktow and Erinsar to increase to the full extent of the capital gains in view of the forthcoming events.

On February 18, 1999, Erinsar (k.) has notified Soktow of its intention to enter into a business combination with Soktow under which Erinsar would obtain the entire ownership of Soktow. In a letter to the Board of Directors of Soktow, Wasker, the President of Erinsar (k.), stated his company's belief, based on information currently available to the company, that the appropriate price for the shares of Soktow held by the public for the purposed such transaction is $14 in cash per share. This price would be a premium of 21 percent over the February 17th closing price of Soktow's shares on the NYSE. Erinsar is currently examining various ways of implementing this decision. The net equity per share was $12. Erinsar held 57 percent of Soktow's shares and the offer will cost Erinsar, assuming full acquisition, about $98M, for about 7,012,967 shares. The market capitalization was about \$200M and the shares outstanding were 15.9M.

Here again, it is not a straightforward offer to the minority shareholders, Erinsar does not want to be liable for a lawsuit if they change their mind. They announced that they intended to enter into a 'business combination' with Soktow. In Furolias' case they mandated Osttowar to give a fairness opinion on the financial valuation of Furolias and not of its intrinsic valuation, in Soktow's case they are looking for a 'combination'. We know already that Istovius, the Chairman of Erinsar, is a genius in "combinatorics", so this announcement should put a red signal to sophisticated shareholders, who unfortunately are very few, and comprise mainly of old-timers as Astossg or Karisios, who can read between the lines. Anyhow, the press release of Erinsar has achieved its goal and the Israeli newspapers understood that the shareholders of Soktow were offered a fantastic offer by Erinsar to purchase their shares above the current price in the stock market and the net equity per share.

We remember how Karisios has foreseen in 1996 that this will come and that the Soktow shareholders will be wronged as Furolias' shareholders were. He wrote it to the US SEC but to no avail. Nevertheless, the Board of Directors of Soktow announced that they would consider Erinsar's offer. We remember that Istovius was also the Chairman of the Board of Soktow, so Istovius had to consider his own offer, but he could of course consult Gorekius, the CEO of Nalodo, the parent company of Erinsar and Soktow, who had gained a lot of experience in Furolias' case in weighing an offer he has made to himself. In Erinsar's case, however, the troublemaker was Astossg, the former CEO of Soktow, who did not keep quiet and was interviewed in all the newspapers about the wrongdoing of Istovius and Erinsar. Normally, in order to analyze the fairness of the offer one should analyze the fundamentals of the company, although fundamentals have a very limited impact as we have learned in all the cases of this book. It is suggested that in every case, and especially in cases of companies with a doubtful ethical record, the ethics of the management and affiliated shareholders should be examined as it has a much stronger impact than the fundamentals. But ethics is not a tangible value, and analysts prefer by far hardware over software.

The price history of Soktow's shares was not brilliant. In fact after 1993, the year when the price reached \$45 (in reverse split terms) and Istovius sold his shares at a high profit, the high price was 25.63 in 1994, 17.5 in 1995, 15.38 in 1996, 9.13 in 1997 and 13.00 in 1998. The low price was much lower, 8.75, 8.75, 5.63, 6.25, 6.25 respectively. So, on face value, to the tired shareholder who has lost hope in the recovery of the price, the offer was excellent, twice as much as the average low price in the last five years, and even more than the average high price in the last three years. The institutional ownership was only 12%, Erinsar had effectively full control of Soktow with 57% of its shares, so what could the individual shareholders holding 30% of the shares do. The Price to Sales ratio was 0.62, the Price to Book was 0.95 and the Price to Earnings (low) in the last five years was 6.78. The five years Growth Rate was only 6.5%, the Net Profit Margin in 1998 was only 0.23% and the five year average was 4.27%. The Return of Equity five year average was 6.79%. Only two brokers recommended hold to the shares and no one gave a strong buy or moderate buy rating. Sales of Soktow were about $300M since 1995, and earnings per share deteriorated from $0.91 in 1995 to $0.51 in 1996, $0.04 in 1997, and $-0.11 in the first three quarters of 1998. No, definitely, the results of the company were not brilliant, and the shareholders who persisted with the company over the years were disillusioned, especially due to the fact that in the meantime the stock market had a very high return on investment. Actually, we learn from the Internet that some of the shareholders were very happy with the sale of Soktow's activities as it doubled Soktow's shares price from about $6-7 prior to the September 1998 deal to $12 and more since that date.

But this was not the issue! Soktow's shares did not reflect any longer the performance of the company, as almost all of its activities were sold and the company reflected mostly the amount of cash it had from the sale of the activities, $370M, less taxes and expenses. The relevant question was therefore what is the proper valuation of Soktow, consisting of a cash pot of hundreds of millions of dollars, a manufacturing activity in Israel and some small activities. This was exactly the case of Erinsar as well, as it held the revenues of $228M from the sale of Deon, 57 percent of the $370M revenues from the sale of Soktow's activities, and some small other operations. From these sums of $600M (!) had to be deducted taxes and expenses, with many divergent opinions on their amount. So, what is the true valuation of Soktow's and Erinsar's cash pot? That is the question, and it depends on 'to be or not to be an affiliated shareholder'. The following events will prove that there is a different valuation of a cash pot to affiliated shareholders and to non-affiliated shareholders. We remember that for the unaffiliated shareholders of Loskron and Furolias, their company was worth almost nothing and was on the verge of bankruptcy, but for the affiliated they had very good prospects and could turn to profitability once they got rid of the other shareholders. The 1994 two former sagas of Loskron and Furolias were repeated in 1999 with the two new sagas of Erinsar and Soktow, showing that there are 'deux poids et deux mesures' and a differential valuation for the same shares, depending on who owned them!

But this case will be meaningless without referring to the personal implications of his heroes. On September 16, 1998, after concluding the basic deal of the sale of Soktow's activities, Istovius is interviewed in Gassan:

'Istovius, president of the Erinsar group, currently also serves as president of Soktow. Astossg, who served as president in recent years, resigned due to differences of opinion concerning the company's future. Astossg believed in Soktow's struggling independently against international k. .... giants. Istovius and most management members thought it better to sell off the operations, and this opinion ultimately prevailed, forcing Astossg to resign...

(Gassan) - Soktow is due to receive $375 million in cash when the deal is closed in November. Have you already decided what to do with the money?

(Istovius) – 'If and when the deal is closed, we shall come to Soktow's shareholders and offer alternatives. Obviously, some of the money will be earmarked for distributing a dividend which will compensate shareholders, and some will go to new investments. It is still not clear how much will go where. We have not decided what to invest the money in, but in my personal opinion, the field of ... in general and k. technology in particular are faster growing fields, carrying great potential. Soktow's product lines are considered mature with relatively low growth rates, and I therefore think we should move in the direction of innovative technologies related to less mature fields. There are lots of ... sector fields not yet penetrated by technology. That is where high growth rates can be achieved and added value can be gained for the company."

This article indicates that legally, if within a few months, the money will be invested in real estate in Hungary, Istovius cannot be sued, because he expressed his own opinion and not the opinion of Nalodo, his main shareholder. Gorekius in the meantime did not express his opinions, and he was the ultimate boss, or rather his shareholders, the Durtem group. But even textually Istovius fulfilled his promises as some, a very small part, of the cash was distributed as dividends, and some, a very small part, was invested in the k. field. Istovius did not mention the order of magnitude of the cash proceeds and if in fact almost all the cash pot was to be invested in real estate by a new shareholder, who bought Erinsar's shares from Nalodo, how could Istovius change the course of action? He is only an employee. So, he receives a few more millions and leaves the arena, Gorekius receives a few more millions and retire, and the only ones to pay for the 'pot casse' of the cash pot are as usual the minority shareholders who receive only a fraction of the proceeds.

The illusionary prospects of class actions can be illustrated from the response of Erinsar and Soktow to the Israeli District Court on February 21, 1999 on a class action against them (one of many to follow) amounting to eight million shekels. The grounds for the application was suppression of information relating to the sale of part of Soktow's assets. The companies claim that, since no share sale deal was involved, there was no obligation to report that negotiations were being conducted for the sale of the assets, and the Securities

Authority had exempted Soktow from submitting regular reports. The company that submitted a class action against Erinsar, Soktow, Istovius and others, claimed that there is grave suspicion that insider information about the negotiations had been used by important insiders at Erinsar and Soktow, and that the information was divulged to some of the option warrant holders, and some of the options were exercised in consequence. A few days later another class action was submitted to the Tel Aviv Regional Court by two private investors against Nalodo, Erinsar and Istovius. The investors held options of Erinsar and requested from the Court to compensate options holders, who sold them in August 1998, in the amount of 6.5 million shekels. The price of the options was almost nil as there was no benefit from exercising them. However, 87,000 options were exercised by Istovius, thus incurring a 'nominal' loss for Istovius of 216,000 shekels. But a few days later, in the beginning of September, the agreement of the sale of Soktow assets was disclosed and Istovius and the other options holders made a large benefit, because they took advantage from their insider information.

Here again we see the insurmountable gap between law and ethics. The managers who probably knew of the negotiations exercised the warrants on their last day when their exercise price was much higher than the market price, a totally illogical conduct if they did not know that the shares' price was going to increase substantially within a few days. Since the company's managers can be accused of wrongdoing but not of stupidity, we can assume that they did so because they held insider information that the other shareholders did not possess and therefore did not exercise the options. Those managers now hide behind the legalistic excuse that they had to divulge only negotiations on sale of shares and not sale of almost all Soktow's assets, which has a much more substantial implication on the shares' price than the sale of some shares. There is no doubt that the managers' conduct was totally wrong from an ethical point of view as they discriminated against the minority shareholders and made unethical gains from insider information. But probably they were right legally. Even if they will not be convicted, the old Jewish saying could apply to their conduct: It may be kosher, but it still stinks...

The managers of the group persevere in walking on the edge of the law. In Furolias' case they did not disclose to the shareholders assembly that was held on June 27, 1994 that the takeover by Erinsar was imminent, as it was to be published only on July 1. They were not required to do so by law, so they did not, and the minority shareholders received the takeover as a 'fait accompli'. In Soktow's case they also act strictly by the law, and those legal actions cause them outstanding profits, while causing the other shareholders substantial losses. The Erinsar case will evolve more and more over this edge, thus proving that you can be totally right legally while being totally wrong ethically. The shareholders meeting of Soktow held on November 20, 1998 on the sale of Soktow's assets was approved by 99.4 percent and probably Astossg was the only one who disapproved. The shareholders could not have guessed of the salami tactics that were about to deprive them of most of the benefits from the transactions. Speaking of the finalization of the transaction between Gosstik and Soktow, Istovius said that it 'is a testament to the talent and dedication of the Soktow employees.' And he could not chose a better term, as in fact the testament was not a tribute like he meant but the last will or free will of the Soktow's employees and the testament of Soktow, which ceased to exist together with the vision of its founders in order to become a real estate company buying hotels in Hungary.

And on February 22, 1999, we learn in Yarmuk, the largest Israeli newspaper, that Suram, the first CEO of Soktow, who took all the blame of Soktow's shares price collapse in 1985 returned to Israel after staying abroad for 14 years. He lived all those years in the US and his absence enabled all the other directors involved in the collapse, such as Gorekius, the CEO of Nalodo the parent company of Soktow, to maintain that they did not know of the events that preceded the collapse. On the same page of the article on his return we could read a small notice on the cancellation of Suram's course at the Business Administration Faculty of the University of ..., benefiting of substantial financial assistance of the Durtem Group, the holding company of Nalodo, Soktow, Furolias and Erinsar. The subject of Suram's course was 'the prediction of success in start-up companies in the information era'.

The Dean of the Faculty said that it was Suram who initiated the course, but the course was not approved by the teaching committee of the Faculty. 'The subject was considered in the teaching committee and we have decided that the subject was not interesting enough to us on academic ground', he said. We can only wonder what does it take to be of interest to the public or to the academic world, and if troublemakers like Suram were not bound to remain silent even when they came back to Israel and would want to disclose the 'pot aux roses' on the cash pot of Soktow, the first high-tech company to be traded in the US, and the flagship for many years of the high-tech industry in Israel.

On February 18, 1999, 'JusticeReed' wrote on the Internet about the planned takeover of Soktow by Erinsar: 'We are investigating the fairness of the transaction to Soktow's minority public shareholders, and whether Erinsar, the majority shareholder, is treating the minority shareholders fairly. We would like Soktow public shareholders or other persons having information pertaining to the fairness of the proposed transaction to contact this firm at ... For more information about Soktow or about class action lawsuits, please review our web site at Sammel...' This post brought harsh criticism in the thread. Chickenlyttle wrote on February 26: 'As soon as there is some cash on the table, count on the professional parasites to try to shake down the company for some cash with their boilerplate nuisance suits.' In spite of these reactions, the Soktow shareholders did in fact file a class action in the US with Sammel.

On January 12, one month before the company was sold, Erinsar reported of its resolution to issue to its Chairman and CEO, Istovius, 676,709 Stock Options, which constitute 3 percent of the equity at an exercise price of $7 per option. He received a loan from the company to finance the purchase of shares. Subsequently, it was stated in the press that the value of his shares in the three Erinsars was above $10M. This amount is of course after the revenues he got from the sale of shares in the past, including the shares of Soktow. According to the press, he did not have any shares in Soktow, although he acted as CEO of the company. On February 25, Erinsar announced the distribution of a $2 per share cash dividend to Erinsar's shareholders. Erinsar's share was traded at $11.5. The valuation of Istovius' options at the current market price was above $3M. The price of the share fluctuated in 1998 between $7 in March 98, $8 from May to August, and $10-$11 from October onwards.

And on the same day, February 25, 1999 at 17:52, 'le coup de grace', the ultimate exploit. The best move from Nalodo's point of view and the worst for the minority shareholders of Erinsar. 'Nalodo (NASDAQ ...), a leading multinational high-technology holding company, announced an agreement to sell all its holdings in Erinsar (NASDAQ ...) to Ertel, an Israeli public company traded on the Israeli stock market. Nalodo will sell 8,575,448 shares (representing 37.3 percent of Erinsar stock on a fully diluted basis) for a total consideration of $145 million (representing approximately $16.9 per share). Nalodo's income, net after tax, will be approximately $20 million. Dividends received by Nalodo from Erinsar, before the closing, will be deducted from the $145 million payment to Nalodo... Gorekius, Chairman and CEO of Nalodo, said: 'Our sale of the holding in Erinsar follows our strategy to focus Nalodo's future growth primarily in the ... fields'.

Sic transit gloria mundi, the shortest Requiem ever written, Bossuet could not have written a better funeral oration to the jewel of the crown of Nalodo, who was sold following its strategy to focus on new pastures. No mentioning of the glorious past of Soktow and Erinsar, who are expendable for a total consideration of $145M. No mentioning of the employees who were transferred to foreign owners and to a real estate contractor, of the minority shareholders of Erinsar and Soktow who were about to lose hundreds of millions of dollars in valuation, of the government who rescued the company and gave Soktow and Erinsar hundreds of millions of dollars in subsidies without recouping most of it, of the high-tech dream that has evaporated in the mist of the golden vapor emanating from the cash pot of Soktow and Erinsar.

When the companies were in trouble, Gorekius and Istovius called for help from the government, the employees and the minority shareholders. But those groups could be no partners in the cash pot, which has to remain for the exclusive benefit of the affiliated shareholders. Nalodo receives from Ertel a premium of $5.5 per share, a third of the valuation or about $50M, in comparison to the market price of about $11. All the others will have to sell their shares at the current market price or much below it, because of the massive sales, as it will go down in the near future to $8, less than half of what Nalodo received. The new owner of Erinsar will decide to withdraw the offer to purchase the shares of Soktow at $14, and the price will collapse as well to about $7. The other shareholders of Erinsar, holding about 14M shares, 63 percent of the equity, will lose about $100M in comparison to Nalodo, and the other shareholders of Soktow, holding about 7M shares, 43 percent of the equity, will lose about $50M in comparison to Erinsar's offer that was withdrawn and more than $100M in comparison to what Soktow's 'dissident shareholders' think should be the true valuation of Soktow with hundreds of millions of dollars in cash.

The overall loss of the unaffiliated shareholders in comparison to the true valuation of the companies measured by the price that Nalodo got for Erinsar's shares and the cash reserves of Soktow, is therefore at least $200M, a long way from the 'miserable' gain made on Furolias amounting to a few tens of million of dollars. Gorekius, Istovius, Erinsar, Soktow, Nalodo and the Durtem Group have definitely improved their methods in the five years that have elapsed since the Furolias saga, and have succeeded once again in a brilliant deal, unfortunately to the detriment of the minority shareholders.

The valuation of Erinsar at about $400M for Nalodo, based on the share price of $16.9 paid by Ertel, is equivalent to the net cash reserves and other assets of Erinsar and Soktow, although Erinsar has only 57 percent of Soktow. But we will see that Ertel, while controlling only 57 percent of Soktow will induce it to invest in its real estate business at very high valuation, thus controlling effectively the cash pot. Furthermore, Ertel, owned by the real estate entrepreneur Zrontius, paid Erinsar's high valuation only for 37 percent of the shares. The acquisition of more Erinsar shares in the open market (up to about 50 percent of the shares) was made at much lower valuation and shares' price. Ertel purchased on March 1, 1999 about 11 percent of Erinsar's shares for about $25M, at a price of $11.5, which was one third lower than the price paid by Nalodo, but higher than the current market price which fell to about $10, following Erinsar's acquisition. The sellers of the shares were institutions, mainly Boral funds, probably the same ones that have lost so much in Furolias' case, (when will they ever learn...). The same applies to Soktow, as Ertel purchased more shares of Soktow not at the 'high' valuation of $14 as was initially offered, but at prices up to 50 percent less. For Ertel it was therefore an excellent deal, as they controlled $400M, most of it in cash, and paid for it about 40 percent of the cash, a fantastic leverage. They can use and will use this cash of Erinsar and Soktow to purchase from Ertel real estate at very high valuations, increasing even more the gains from this transaction.

Following the unprecedented press campaign that condemned Nalodo, Erinsar, and Gorekius for selling Nalodo's shares to Ertel, Gorekius had to respond. He was interviewed in Yarmuk on March 2, 1999. 'The investors did not like at all, to say the least, that they went to bed with the shares of a company that belonged to the high-tech group Nalodo headed by Gorekius and woke up with shares of a company belonging to a real estate entrepreneur. The surprise, it has to be said, was great: from Gorekius, who is viewed in Israel as the founder of the high-tech industry and the start-ups, it was not expected that he would sell Erinsar to a contractor, even not for a substantial gain. Erinsar becomes now a financial organization, that will be financing the high risk real estate firm of Zrontius in eastern Europe, which is operated by his company Ertel. Zrontius himself has declared at the eve of the transaction that he intends to use the heavy cash pot of Erinsar to continue building hotels and malls abroad. Zrontius, from his side, has done an excellent deal: For $145M he controls cash of $300M, and it is doubtful if he would have succeeded in raising this amount in the stock exchange or from banks. Nalodo has gained a respectable capital gain of $20M. And who does not participate in this festivity? The shareholders of course. At least until now...

In this interview Gorekius maintains that the source of the attacks on him is emotions and not a rational economic thought.
Yarmuk - The shareholders are angry at you, and yesterday they expressed their anger by a wave of sales, that brought down the share's price by 13 percent.
Gorekius – What happened yesterday in the market is a human uncalculated response, and I think that those who sold yesterday will regret it. If we speak about emotions, then I, who separate from a company that I have built, have of course more emotions, but I have obligations to the welfare of the company, to its customers and to its shareholders, that I think have benefited from us. We have to look at what happened to the price of the share: it doubled over the last year and we also promised before the sale a dividend of $2 per share. So why are they angry?'
Yarmuk - Maybe the anger is that the shareholders have bought a high-tech company, and they found out that they have received a real estate company. When the activities of Erinsar were sold to the foreign companies you declared in the press that the cash is earmarked for new investments and dividends?
Gorekius – That I didn't say. (It was said by Istovius, the CEO of Erinsar – Yarmuk) Nalodo has declared once and again that its strategic field of activities is … It is very easy to criticize, but what did I sell? Not a company but a cash pot. People distort the picture. The high-tech business of Erinsar

was sold prior to that to Gosstik and Priam. Now we took only our part of the cash and we shall use it to enter into the fields that Nalodo believe that are good for it... Those who criticize the deal are those who thought that they will gain more from speculation and thought that they will be bought out at another price.'

It is amazing how the cases may be different but they have always the same rule, the GI rule. We remember how the shareholders of Loskron were called speculators when they held their shares for years and years and lost all their money, Karisios was also called a speculator in the case of Furolias after holding his shares for three years and not selling them in order not to use insider information. Here, Gorekius rides again on this horse. It is always worthwhile to blame your adversaries of your own defaults: 'The evil person attributes to another person or group exactly the deed he himself is about to perform, a deed which victimizes this person or group (Arendt, 1979; Young-Bruehl, 1996)' But this is not enough for Gorekius, he wants to act as the benefactor of the shareholders. He probably knows that it is not true and they are going to lose most of their investment, as this case proves, but he says that those who sold yesterday will be sorry. Why, because the shares' price collapsed even further subsequently? Gorekius has 'obligations to the welfare of the company, to its customers and to its shareholders, that I think have benefited from us.' Gorekius, the ultimate benefactor, Mother Theresa of the poor and oppressed, must have only one interest: to make the partners of the company – the customers and the shareholders – benefit from his actions, as he has obligations to the welfare of the company, which he has sold as a cash pot in a flagrant discrimination of the unaffiliated shareholders, keeping up appearances as their benefactor. (see 'Evil agents cast themselves in the role of a benefactor, Lifton, 1989; James, 1984).

Only Gorekius does not get emotional, after all why not, he never loses, it is only the minority shareholders in his companies that lose. He told Karisios that some of his best friends have lost money in his companies and only Karisios complained, but this time he cannot say it anymore. The largest economic institutions are going to sue him in November 1999, including some of his best friends. But pressed to the wall, Gorekius is even willing to blame Istovius indirectly of making a hasty announcement after the sale of Erinsar's assets that the proceeds will be invested in k. high-tech and distributed in dividends. Actually, Istovius was much more cautious in his interview, as was analyzed before, but even this 'mistake' could not be made by Gorekius, who really did not make any mistake in his interviews, dealings, press releases, announcements to the SEC and to the shareholders, and so on. This brings us to the infallibility law of the Gs (Gorekiuses), they never lose, they never make mistakes, all the others do. Gs do not disclose all the facts, they are silent about their intentions, they pronounce statements that have double meanings, but throughout all these cases there is not a single event in which the Gs made a mistake that could make them seriously liable for a lawsuit. They may be sued, but the lawsuits are unfortunately ungrounded legally. Ultimately, the Gs' conduct is completely unethical, but they go on the verge of the law, surrounded by the best lawyers and PR agencies, and the law, the press, the public, the shareholders, even the government, cannot touch them.

On February 22 we read that Obsol, the head of one of the Israeli government's organizations, announced that Soktow was delaying payment of royalties to the government on the transaction of the sale of Soktow's activities to Gosstik. She wrote to Soktow that the delays in payments were not concurrent with the agreements reached with the government on the transaction. For weeks, Obsol was going to conduct a crusade against Soktow, Nalodo and Erinsar, especially after the sale of Erinsar, stating that they did not fulfill their obligations to restitute to the government part of the incentives that were given to Soktow and were due after it sold its technology. She even started her own investigation and sent an investigator, Blon on March 12 1999, to Karisios and Poftrim, and later on to Astossg, to receive testimony, as she maintained that Erinsar has to commit to invest $200M in high-tech as a condition to the approval of the sale of Erinsar. On March 10, 1999 in an interview with the Horton newspaper, Obsol dared for the first time to attack personally the ruling family who owns Nalodo and say that they do nothing to restitute to the government what is owed to it - $197M. She even received full backup from the minister in charge of her organization who said: 'The struggle of Obsol against the Nalodo Group is right'.

But Nalodo stood firm. On an announcement on the same day they maintained: 'We wish to state that we reject all of these arguments against Nalodo, Erinsar and Soktow. Based upon a legal opinion received by Nalodo, it is clear that our actions were faultless and that no approval of the ... organization for the said transaction is required.' Obsol replies 'you cannot base everything on the law. In the relations between the government and the companies there should be ethics and trust.' But the campaign proved to be 'much ado about nothing' and a compromise was reached that seemed to many observers to the detriment of the government. On March 25 it was reported in Horton that the group Nalodo/Erinsar has reached an agreement with the government and Obsol. It was stated that Nalodo and Erinsar will not change their business policy and will continue in the future to invest in k. Ertel committed to invest in high-tech an amount of $45M, without submitting a detailed schedule. $15M from this amount will be invested in k. in Erinsar. Nalodo committed to invest $150M in Israeli high-tech. From this amount Gosstik and Priam will invest $50M. No schedules, just general statements, no money received by the government, just a letter of intention stating what is in the mandate of Nalodo anyhow - to invest in the high-tech industry. But what about the huge amounts invested by the government in Soktow, how will they ever be recouped? Once again, as in the case of Suram, of Karisios, of Astossg, and of many others, the protest was silenced and the Nalodo group was able to consummate the deal with huge profits to themselves, without sharing them with the shareholders and the government.

After the sale of Erinsar and Soktow to a real estate contractor, Poftrim had at last made his point: that Erinsar perceived all over the years Soktow as a financial acquisition and never intended to invest in the company and to develop it. In an interview in Gassan on March 11 he says: "'Every step Erinsar took in Soktow since buying it ten years ago was made with the prior intention of selling the company,' says Poftrim, who was president of Soktow from 1988 to 1995. He was commenting on the agreement for the sale of Soktow's parent company Erinsar to developer Zrontius. 'Erinsar never believed in Soktow,' Poftrim added, claiming that Nalodo pressured Erinsar into buying Soktow shares because it suffered from poor cash flow in that period. He went on to say that Erinsar and Nalodo entered into a series of apparently unconnected transactions that in fact 'were made with the clear intention of benefiting Nalodo and co., at Soktow's expense.' Poftrim claims that, in the agreement Erinsar signed to buy Soktow's shares, it undertook to invest $10 million in the company, but had not done so. He also commented on the acquisition of Deon by Erinsar in1994. He called this purchase 'theft under Soktow's nose', and claimed, 'they stated that Deon wanted Erinsar rather than Soktow to buy it. ..., then Deon chairman and president, told me that Istovius made the deal conditional on the company being sold to Erinsar, and not to Soktow.'

(Gassan) – But what do you want, Istovius bought a company at $70M and sold it to Gosstik at $230M. What's wrong with that?
Poftrim – He is really superman. It is a fantastic profit, and net, as in Deon there was no participation of the government. From here on it is my speculation: Deon was sold at a very high price. But the divisions of ... together were sold by Soktow to the same company at $100M. What happened here? The proportions are not at all adequate. I don't know how to explain it, unless the brain goes in a direction of not telling the truth to the public or concealing something. Erinsar holds 60 percent of Soktow. From money that goes to Soktow, Erinsar receives only 60 percent. From money that goes to Erinsar, they receive 100 percent. Another facet is that part of the money that was paid to Soktow has to go back to the government that invested a lot of money in Soktow. But in Deon there is no government money. This is my estimate that is not backed by facts. So there is a problem. Someone will come and will sue them in a class action and they will have to testify and we shall see what they will say.

Poftrim says that the Israeli government has to receive back from Soktow about $150M but they received 30 and were satisfied with it. Obsol says that it should be $76M, and Istovius gets angry and says: 'he would better not tell figures that he does not understand. So he shouldn't speak'. 'I even don't talk about the loans of the banks to Soktow that were foregone' says Poftrim 'and all because of the charm of Gorekius. He and Istovius know how to charm people, and it is part of the problem in this country, that with connections and charm you can achieve things. I don't know why Obsol has not received the money she was entitled to. She said once 'they put me before a fait accompli'. It isn't true. They could have demanded it. But what, they gave up. They always give up. Why not? It is the money of Obsol? It is the money of the state.'

Poftrim knows that as far as personal communications, convincing capabilities and charm is concerned, Gorekius will always win. He remembers well how once, as CEO of Soktow, he set a meeting with a reporter of a large newspaper and within a quarter of an hour Gorekius knew about it and phoned him: don't talk. No, he has not developed a theory of personal conspiracy, he brings this story as an example of the excellent connections that Gorekius has with the media. 'An excellent manager, Gorekius, winner of the prize of ...', he says in irony, and immediately enumerates a series of business failures of Gorekius. He says that Gorekius charms everyone in casual meetings, but if one knows him well, he sees that 'Gorekius is a sticky charmer'. Istovius, also interviewed, would refrain from personal remarks and would only mention 'the arrogant management of Soktow' when Poftrim was the CFO.

(Poftrim) – 'There is no one who speaks about the shareholders as Gorekius. Like Bibi (Netanyahu) who yells all the time 'security, security', Gorekius yells all the time 'shareholders, shareholders'. If this is so, why when there is a company without operational activities, and with a lot of cash, he does not distribute the money to the shareholders? And where is the Israeli SEC? The shareholders are among others pension funds, it is our money, not somebody's else.'... 'From the moment that I saw the vulgarity, the foot with the boot that enters into the company, it was clear that the company is doomed. When I left, I wrote to the Board of Directors 'I think, that the company has started to go in a way without possibility to succeed.'

Erinsar's president, Istovius, said in response, 'This is a malicious smear. I, and everyone who worked with me, tried to obtain as much money as possible in every deal. We conducted independent negotiations and tried to get the most we could. As it happens, I have letters from Gosstik proving that every sale was made separately, and there is no link between them.' Istovius commented on the other claims that 'it's not true that there was pressure on Erinsar to buy Soktow, nor is it true that we didn't believe in Soktow. We bought it because we wanted to transfer from the s. field to the r. field. As to the $10 million, there was an agreement that, if investment proposals were put forward, we would invest such a sum. But no investment plan was ever submitted.'"

No investment plan was ever submitted, no other investor was interested in buying Soktow, we conducted independent negotiations, we have letters proving it, people who sell Erinsar's shares now will regret it, Deon's management did not want to be merged with Soktow and I could not do anything about it, etc. Poftrim, Astossg, and others say one thing, Istovius, Gorekius, and others say the opposite, and between them the minority shareholders are losing most of their investment. But at least, Poftrim was heard, he was interviewed, the shareholders had the opportunity to decide for themselves whom to believe. Poftrim said that Deon's purchase by Erinsar was 'theft under Soktow's nose' and was not sued for libel, probably because Erinsar was afraid of the Pandora's box that will be opened in this trial, and why sue Poftrim, as the ultimate deal was consummated and they had no other plan in their pipeline. They have cashed whatever was cashable, disposed of whatever was to be disposed, the warriors could at last rest.

In a conversation between Karisios and Otwuss, the eminent lawyer and scholar who tried to mediate between Karisios and Istovius in the Furolias case and had a great respect for Karisios' ethics and integrity, Otwuss summarized the changes that have occurred at Erinsar, since Karisios left them and the final strike of selling Erinsar's shares to a real estate contractor, by saying that 'Erinsar has declined from Karisios to Zrontius'. But who is Zrontius, is he a worthy heir of Erinsar's glory, or a speculator who seized a fantastic opportunity to control a cash pot at an attractive price. Here is what he says in an interview in the same day and in the same newspaper: "'I am not going to lay off anybody. We have received a company with an excellent R&D team headed by Wasker. This team can develop glorious companies worth three times more than Soktow was sold. Don't be surprised that in the 2000s Wasker (the president of Erinsar) will found a new high-tech company worth more than Soktow and will sell it again to foreign companies at a high price. Don't forget that at Ertel we have high-tech companies, that some of them are in parallel fields of Soktow's. This will be a more interesting story than high-tech.

We are constructing in Europe 50 malls at a total investment of $1.25 billion in five years. In the hotel business we shall invest more than $400M. In the future I intend to separate the high-tech from the malls and hotels. Two months from now we shall present a detailed plan, and in the meantime there is no reason to panic. In March '96, I bought the company ... at 400 points, and today this is Ertel, traded at 3,100 points. The prices are so low because it is traded in Israel. Abroad it would not have received these low prices. Erinsar too, in its present position, is traded at a price that is much lower than its net equity. Erinsar has announced that it contemplates the possibility (of purchasing the minority shareholders' shares), but has not published an offer to purchase the shares of Soktow. It may be possible to reconsider the proposal, but I cannot promise anything at this stage.'"

Zrontius is very cautious, he does not promise to purchase the shares of the minority shareholders of Soktow at $14, as Erinsar has vaguely announced. It is very important to notice, as Ertel will decide later on not to purchase the shares at $14, to purchase some of them at a much lower market price, and to induce Soktow to invest the cash pot in its real estate companies at a valuation which seemed to Soktow's shareholders exorbitant. The shareholders of Soktow decided therefore to sue Erinsar. On November 3, 1999, Yarmuk reported: "14 institutional investors of Soktow, including the funds of Bank Horshrem, and 10 other shareholders, are suing in the Haifa District Court to force Erinsar to implement the offer to purchase the Soktow shares at $14 a share or alternatively to compensate them on the damages incurred by them following Erinsar's decision. The lawsuit is addressed also against Nalodo, the group controlling Ertel, Zrontius, and officers and directors of those companies. Among them … Gorekius, Istovius, …

The institutional shareholders are requesting in parallel that the court will treat their suit as a class action on behalf of all who held Soktow's shares since 6.9.99. The plaintiffs maintain, with the lawyers …, that the minority shareholders of Soktow are extremely and continuously discriminated against by the group that controls the company and its directors. They ask that Erinsar be forced to make the purchase of the shares. Alternatively, they ask the court to prevent the fulfillment of the transactions between the companies controlled by Zrontius and Soktow to acquire hotels abroad and entertainment centers in … After the change of control, say the plaintiffs, Zrontius has caused to Erinsar to withdraw from the offer to purchase the shares of Soktow. Companies controlled by Zrontius have concluded agreements with Soktow selling to Soktow for $196M, hotels with a valuation much lower than this amount. In view of these acts, say the plaintiffs, the value of Soktow's shares has dropped by 45 percent since February from $13.25 to $7.25 per share."

In order to understand how the shareholders decided to sue Erinsar, we have to analyze the sequence of events that led to this lawsuit. On March 3, it was published that an independent committee formed by Soktow will consider the offer to buy out the shares of the minority shareholders. And who was appointed to this task as financial advisors on this matter to assist the independent committee of Soktow? Our old friends Osttowar who advised the independent committee of Furolias on the fairness of the price offered by Erinsar to the minority shareholders of Furolias. Plus ca change plus c'est la meme chose... Soktow's press release on the same day clarifies that the independent committee consists exclusively of outside directors (in Israel the 'independent directors' were called 'outside directors', but the new

Companies Law changed the terminology to independent, which will probably have a very favorable impact on their independence...). This committee has issued a letter dated March 2 on its willingness to begin negotiations regarding the proposal. And most importantly, Soktow's release is very explicit in the fact that on February 18, they received a letter from Erinsar that proposed to acquire all of the outstanding shares of Soktow held by the public for cash consideration. No vague intention, but a firm proposal that is on Soktow's option to accept or reject. We remember that Astossg thought that the price of $14 was too low, and ultimately he sued Soktow and Erinsar in a class action suit, requesting a price two times higher - $28.

In parallel, on March 2, Ertel also acquired 2,221,151 shares of Erinsar, 10.2 percent of the total shares, at an average price of $11.6 per share. The shareholders who sold their shares, sold them at a discount of 41 percent compared to the price of $16.9 that Nalodo received from Ertel a few days earlier on February 25th for the sale of 8,575,448 shares, 39.39 percent of the total shares. Ertel now held 49.6 percent of the shares according to Erinsar's press release. The selling shareholders, mainly institutions that ultimately sell shares of the investors in their funds and not their own shares, were happy to get rid of the shares at a discount of 41 percent. The precedents of Furolias and many other cases with the Nalodo group proved to them that in those cases the sooner you sell your shares, you cut down your losses at a 'reasonable discount'. They were proved right as the shares' price fell even further to $8, once Zrontius had 50 percent of Erinsar's shares and effectively full control of the company and did not need to buy the remainder of the shares. Erinsar's and Soktow's main purpose now was to be given a financial leverage to buy through their funds Ertel's real estate assets at an attractive valuation. This could be done when you have 50 percent of Erinsar's shares and 57 percent of Soktow's shares and you do not need 100 percent of the shares. The institutions, that knew that, were therefore happy to get $11.6 per share. Other institutions, some of them from the Durtem group itself, that did not succeed in selling their shares of Erinsar to Ertel, approached on March 15, with their lawyers the directors of Erinsar in a request for a substantial distribution of dividends to the shareholders of Erinsar. It was reported that Zrontius did not agree to purchase the shares of these institutions, probably because he did not need to increase his ownership of Erinsar.

Why the institutions did not get rid of their Soktow shares is an open question. Maybe they were confident that Erinsar will stand firm on its 'obligation' to purchase the shares at $14, maybe there was a smoke curtain with sous-entendus that induced them to believe so. Probably Erinsar has issued its 'offer' a week before the sale of the shares to Ertel, in order to have quiet on Soktow's front, as Nalodo was afraid of the implications of the sale of its shares to Ertel. One thing is certain, the two moves were coordinated as nothing is casual in Nalodo's group and everything is concerted meticulously, as we have seen in the Furolias case and in this case previously. But the facts are that Erinsar did not give a firm commitment (they notified Soktow of their intention to enter into a business combination…), Soktow did not agree to it, there were shareholders like Astossg who thought that the $14 price was unfair and submitted a class action, and most of all Zrontius said explicitly that he is not committed to stand behind Erinsar's 'promise'. The institutions made a mistake, because Nalodo, Ertel, Loskron, and most of the affiliated shareholders in Israel and France succeed inexorably to have their way, and they should have sold their Soktow's shares at a loss. But maybe this time, as some of the plaintiffs are important organizations, they will succeed in recouping some of their losses. If they were just individual shareholders they had no chance whatsoever.

In the meantime, Nalodo's shares' price has increased in half a year from about $11 to $17.75. Gorekius and Istovius have succeeded once again to make out of a strategic failure, this time of Soktow and Erinsar as a few years ago of Furolias, a financial success. Could it be that precisely at this moment, their main shareholders, the Durtem group, decided that they were expendable? In the same year, 1999, Gorekius retired on November 1 1999, with a golden cushion of $10M to attenuate the fall, and Istovius received a few million dollars more as he found himself out of job from Soktow and Erinsar (k.), and even in the two other companies of the old Erinsar there were two CEOs who took care of the day to day operations. Istovius left the Group a few months later and resigned from the Boards of Directors of the other companies of Erinsar. In a meeting with Karisios, he even complained that he did not receive a generous farewell bonus as Gorekius did, and he was very annoyed about that…

The 'servants' have served well their masters, and they became expendable at the apotheosis of their financial exploits. Besides, someone had to take the heat of the public outcry on these exploits that made their masters richer by a few hundred million dollars more. No one mentioned the organization or the family of the owners of Nalodo, either it was too remote or they did not want to mess with the richest family in Israel. The family continued to keep its ethical image, being the benefactors of many welfare and cultural organizations, and continuing to assist in financing the Business Administration faculty of the University of …, that was not interested in the courses of Suram, the former CEO of Soktow. The Israeli experience showed that troublemakers were always punished, but good and obedient servants were always rewarded by some remnants of the quarry.

One of the more serious allegations was the allegation made by Poftrim in his article, which could open a Pandora's box, but was very difficult to prove. We remember that Erinsar finalized its sale of Deon on April 9, 1998 for the fantastic price of $228M. Erinsar bought the company three and a half years ago for $70M and did not manage to make the turnaround that it expected. Everyone was astonished from the fantastic price that gave Erinsar a net profit of $103M. Even Karisios congratulated Istovius on his astuteness in negotiations with one of the largest companies in the world, Gosstik. The surprise was even greater when Erinsar announced that it has sold to the same company its divisions of ... at Soktow for the very low price of $100M. The other division of Soktow was sold to Priam at a price of $275M, although the difference in valuation between the two divisions was not so exorbitant. Was it possible to prove that the real price for Soktow's divisions sold to Gosstik should be much higher, and that it was done in concert with Gosstik in order to divert part of the remuneration for Soktow's division, tens of millions of dollars, from Soktow, where Erinsar had only 57 percent of the shares, to the remuneration for the sale of Deon, where Erinsar held 100 percent of the shares, without reimbursing to the government monies that it owed them? Here again, if this was true, the gain to Erinsar was of tens of millions of dollars to the detriment of the minority shareholders of Soktow. But the potential benefit to the minority shareholders of Erinsar from this alleged diversion of funds from Soktow to Erinsar, which will be elaborated in the lawsuits, was only ephemeral as Erinsar's minority shareholders joined their friends of Soktow, a few months later, as both groups have been treated with the same methods. Shareholders of Soktow and Erinsar did not unite, as the proletariat, but both groups sued Nalodo, Soktow and Erinsar and their executives and officers separately.

Karisios has retained one share of Furolias, the three Erinsar's and Nalodo. As a holder of a share in his name he was entitled to receive all the press releases, the announcements to the SEC and the stock exchange, the financial statements and so on. As he ceased to receive curiously enough the releases from Erinsar, although he could gather their summaries from the Internet and the press, he phoned Otto, one of his former employees and today CFO and company secretary of Erinsar. Otto said at first that they were not bound to do so, but after receiving a formal letter from Karisios, mentioning his rights as a shareholder of Erinsar to receive the announcements, he received all the documentation. And effectively Karisios' hunch was proven correct. Erinsar had never promised to purchase the shares of Soktow; all the articles, the Internet posts and so on were not correct. How is it that miraculously enough all the media was wrong in the same direction; who gave them half truth statements, is easy to understand. The unethical heroes of these cases may induce, imply, give impressions, hint, but when they are bound to give legal documents that might cost them millions of dollars they are very precise in their statements. Karisios knew from his work with the Group, after all he was an insider, that when they give immediate reports to the SEC they are very precise and think of the implication of every word. And when he had in his hand the report, he grinned and understood that the poor shareholders of

Soktow had no chance to win legally the case of forcing Erinsar to purchase the shares of Soktow at $14.

The February 18 report to the SEC was written as follows: 'Erinsar has announced today to Soktow its intention to act for the unification of its business with Soktow, so that Erinsar will achieve the full ownership of Soktow. In a letter to the Board of Directors of Soktow, Wasker, the President of Erinsar, has expressed the belief of Erinsar, based on the information the company has today, that the suitable price for the shares of Soktow held by the public in the respect mentioned above, is $14 in cash per share, a price that reflects a premium of 21 percent, in comparison to the last Closing Price of Soktow's share on February 17, 1999, at the NYSE. Erinsar is examining the different ways for implementing this decision. Erinsar has not received yet any response from Soktow in this matter. Sincerely yours, Otto, CFO and company secretary.'

The report mentioned Erinsar's intention, not its commitment, (we can change tomorrow our intention, we cannot change a commitment or a firm offer), the price of $14 is based on the information the company has today (tomorrow is another day, the company may be sold to Ertel within a week, Zrontius may have other intentions, and of course the information is relevant only for today, in a few months the share might be priced $7, but who cares). And most importantly, Erinsar is only examining different ways to implement this 'limited' decision. (We have not received an answer from Soktow. One of the ways to implement this decision could be to sell within a week our company to a contractor who will buy with our cash pot hotels, entertainment centers and malls).

So what was the reason of issuing such a report, a report that did not commit the company to anything, when it was clear that the company was to be sold? And what was the purpose of this double talk: for the SEC, a vague letter of intention, and for the public – a firm offer to acquire the shares of Soktow at $14? It is obvious, Gorekius and Istovius wanted to convey the impression that Erinsar was going to use its cash pot, even after the sale to Zrontius:
- to distribute dividends, as they decided to a limited extent on the same day of the sale to Zrontius, February 25;
- to invest in k. high-tech as Istovius promised and Zrontius hinted;
- to purchase for about $100M the remaining shares of Soktow from the minority shareholders.

In this way they could achieve public quiet, sort out the problems with the government, and attenuate the outcries of the shareholders of Soktow and Erinsar. From the mythological days of Ulysses' Odyssey to our days the tactics are the same. The Lotus-Eaters offered Ulysses' companions the lotus which made them forget everything and be unmindful of what happens to them. Today, the affiliated shareholders give the small shareholders modern 'lotus' or 'lockshen' in order to put them in lethargic sleep and be unmindful of their schemes.

This report to the SEC, matched with the PR campaign that came subsequently, is a smoke and mirror masterpiece to gain time until everything is finalized and the companies and their executives get the money. We remember how Istovius deceived Karisios with similar vague promises in the Furolias case, this is a pattern, the Istovius' pattern, of distributing vague promises while committing to nothing. They probably counted on that, in a few months, everything will be forgotten, the shareholders will settle with whatever is left, and who cares anyhow, as long as we participate in the quarry and receive some remnants of the bowels of the poor old Erinsar that was slaughtered as an old cow (a cash cow), we can leave for the minority shareholders only the bones. It is too good anyhow for those despicable speculators...

The outcry of the minority shareholders of Soktow gathered momentum. The wrongdoing of Nalodo's group succeeded in rallying very notorious businessmen and large companies who committed the mistake in investing in Soktow and Erinsar shares. They believed in the honorable men – Gorekius and Istovius, backed by the Durtem Group with the impeccable reputation, headed by a family of renowned benefactors to the community. It was impossible that they would defraud them. They heard of course of many dubious events, but as it did not happen to them they did not care too much. They did not want to hear Karisios' admonitions in 1994, after all who is he, a miserable whistle-blower who did not have smoking gun evidence, they did not want to see Astossg's admonitions in the newspapers a year ago, after all who is he, an obscure CEO from Haifa, that does not belong to the elites. But this time, Gorekius and Istovius blatantly betrayed the elites who invested in their companies, and to betray the elites by one of their members, is unheard of, even in Israel. To say the truth, Gorekius and Istovius in the case of Erinsar and Soktow, outperformed themselves. They probably knew that this would be their last exploit, and they acted as in a liquidation sale, they forgot all their normal constraints and appearances, they treated the minority shareholders so extremely, so shamelessly, so flagrantly, that the press, the public, and the shareholders could not pass over as they did in previous cases.

The Horton article of November 4, 1999, summarizes it all. It leaves an impression of deja vu, as the lawsuit repeats almost to the word the arguments and the terms used by Karisios in his synopsis of a lawsuit against Furolias. The methods used by Nalodo, Erinsar, Gorekius, and Istovius are the same, only the order of magnitude of the wrongdoing is much higher, as the gamblers in the casino, the true speculators have raised their stakes for the whole jackpot. This time they played 'le tout pour le tout' and nobody was there to tell them 'rien ne va plus': "Rarely are gathered entities so different as, ... , the University of ..., in one lawsuit. Rarely find themselves businessmen like ...., Gorekius, ... sued by Kamtov Industries. The man who succeeded to make this extraordinary event is Zrontius. The affair of the acquisition of the control of Erinsar by Zrontius and the affiliated transactions that ensued, has aroused the fury that gathered brokers from Ehad Haam street and educational organizations. Those bodies have filed yesterday a lawsuit in order to force Erinsar to fulfill its offer to purchase the shares of Soktow at a price of $14, or force Soktow or Erinsar, Gorekius and Istovius to purchase the shares of Soktow that these bodies own. Alternatively, the plaintiffs ask from the court to forbid transactions between interested parties as they are opposed to the interest of the company and to force Soktow, or the shareholders who control it, to restitute monies that were paid for these transactions. These requests are submitted as shareholders' derivative actions on behalf of Soktow, as the plaintiffs maintain that directors of Soktow and the board of directors of Soktow have cooperated in the acts of loot of the assets of Soktow in favor of Zrontius and in the purchase of the shares of Erinsar that were owned by Istovius, which are to the detriment of Soktow.

The plaintiffs name Zrontius as 'a looter of companies and a dangerous bull'. They quote the verdict of the Supreme Court in the affair of the companies that were managed by Zrontius and went bankrupt in the attempt to build the town of ...: 'The deriving of an enormous amount from the cash of the company to the pockets of four entrepreneurs (including Zrontius) in return for a shameful agreement for non-competition between the entrepreneurs and the competition. The companies borrowed money to pay the entrepreneurs and their indebtedness became enormous until they were liquidated.' The plaintiffs attack personally Zrontius, and it is surprising to see Kamtov attacking frontally Gorekius, ... in their conduct in the affair. The plaintiffs who are represented by Naran refer to the conduct of Nalodo, Gorekius and Istovius as the primary cause of the mischief in the Zrontius affair. They maintain that Nalodo and their directors have infringed their fiduciary duty to the shareholders when they sold the assets of Erinsar, held the proceeds in cash and chose not to distribute the monies to the shareholders, while not making any adequate use of the proceeds. Naran states that Erinsar and its subsidiary Soktow did not distribute dividends, and only after they sold the cash to Zrontius 'has Nalodo agreed to throw a bone to the unaffiliated shareholders of Erinsar that distributed $44M as dividends, while not distributing any dividends in Soktow from the equity surplus.'

... Istovius has sold to Soktow, after the transfer of control of Erinsar to Zrontius, all the Erinsar shares that he owned at the price of $10 per share, which was 22 percent higher than the two months average price since the transfer of control, for a total consideration of $8.8M. The plaintiffs state that the transaction with Istovius was deprived from any economical justification or business logic for Soktow. The price of Erinsar has not reached afterwards the selling price of Istovius' shares, thus causing to Soktow a loss of five million shekels. The plaintiffs state that Nalodo and Istovius have taken advantage of their rights as shareholders in receiving cash for their shares, while the other shareholders of Soktow have not received cash and instead have received Zrontius as controlling owner... Nalodo and its Board, Gorekius, Istovius, ... have ignored, in their pursuit of the maximum consideration for their rights in Erinsar and Soktow, their obligations to the minority shareholders and have abandoned them.

A large part of the lawsuit is dedicated to the shares purchase offer of Soktow shares that was cancelled. Soktow has published in February '99, which is during the time period that Nalodo negotiated its transaction with Ertel, a release stating that it has received a proposal from Erinsar to purchase the shares held by the public at a price of $14. The plaintiffs state that in a release issued in this matter by Erinsar, it was not mentioned that there are negotiations to sell the control of the company. Nalodo and its Board have prohibited, according to the lawsuit, the implementation of the purchase offer in the time period between the signature of the sale contract with Zrontius and its closing. Nalodo and the directors in the Boards of Erinsar and Soktow have prevented the implementation of the offer although they knew that the companies will receive a controlling party that covets the cash of Erinsar, and although the Funds management company Arriom has approached them in writing and warned them that Zrontius will act to cancel the offer after obtaining control. The plaintiffs state that Nalodo and its directors were aware of the danger but in this case also they preferred the interests of Nalodo and its officers over the interests of the public shareholders. Nalodo has not forced Zrontius and Ertel to maintain the offer, but they knew how to request from Ertel to hold harmless Nalodo and its directors and executives against any lawsuit that might ensue from not fulfilling the offer.

Nalodo did not sin only by not forcing Ertel to fulfill the offer of Erinsar. Officers of Nalodo continue to represent the fulfillment of the offer. Thus, the company secretary of Nalodo said in an interview that the controlling party of Soktow has provided also for the minority shareholders and not only to themselves, and in the contract to sell the controlling shares of Erinsar they forced Zrontius to fulfill the offer to purchase Soktow shares at $14... For seven months it was represented that the offer is still valid until September 1999 the outside directors of Erinsar decided that 'the financial situation and exposure, based mainly on the outstanding obligations and others that are related to the sale of Soktow's assets in '98, bring them to the conclusion that the offer will not benefit the shareholders of Erinsar.' This announcement brought about the collapse of Soktow's shares, but the Board of Soktow, that was changed after the change of control, has stated on the following day that they do not intend to do anything in order to force the offer although the announcement was to the flagrant detriment of Soktow... The Board of Soktow has preferred blatantly the interests of Zrontius over those of the minority shareholders, in a flagrant breach of their fiduciary duty to the company and its shareholders.

The justification given by Erinsar for the withdrawal of their offer are ridiculous and show lack of good faith in view of what was to take place two days afterwards. Erinsar and Soktow have announced an agreement by which Soktow will purchase from Ertel its hotel business and entertainment and mall center in .... Erinsar has announced that they intend to buy malls of Ertel in Europe. The plaintiffs ridicule the allegation of Erinsar that they are afraid of the purchase of the shares of Soktow, which are almost exclusively a cash pot without activities, but is not afraid from the risk of purchasing malls in Eastern Europe that Zrontius expressed specifically on their risk: 'The US investment bankers come to you with nice suits and perfect English and say: if you want money no problem, the price is Libor + 20 percent.' 'In the running of Zrontius to the cashpot of Soktow, he did not have time to find a suitable excuse', write the plaintiffs. They say that in the analysis of the financial statements of Ertel they found out that Ertel financed the acquisition of Erinsar from Nalodo by a credit line from two banks... 'The transactions that Soktow is planning to be involved in are extremely dangerous in overexagerated valuations, that are in complete detriment to Soktow.'

The plaintiffs are clarifying the nature of the transactions between Soktow and Ertel. Unlike it was published, Soktow will not buy hotel assets but shares of the companies that own them. It means that Soktow takes the loans that finance the purchase of the loans in Europe. The plaintiffs state that Soktow is buying only a portion from a company whose equity is $4.4M, in return of $154M, which is 35 times its equity. Those businesses have generated revenues of $5M in 1998, thus offering the hotel business to Soktow at a multiple of 30 of the revenues, an unheard of multiple which is unthinkable. The plaintiffs have found that the profit before financing expenses that was generated by the hotel businesses in '98 was $24K. 'In other words, Soktow will need 6,400 years to return to itself the investment in acquiring those businesses.' The plaintiffs remind that Ertel or its parent company have never succeeded in selling the hotels to a third party with a reasonable profit... Although those transactions were not in the normal course of business, and although most of the directors have a personal interest on them, Zrontius and other officers have decided not to convene a shareholders' meeting of Soktow, as requested by law, and overlooked the legal obligation to receive the approval of at least one third of the unaffiliated shareholders."

This article and the lawsuit of the 24 plaintiffs, some of them like Kamtov are among the largest organizations in Israel, and some of them like Arriom were founded by banks that have already lost money in the Furolias affair, are edifying and summarizing the whole dilemma of addressing to the Court. This trial if it ever takes place will last for three to five years, after appeals and so on, it can even reach 10 years. By then, the companies may no longer have cash, yet the lawyers will have plenty of cash, because it is going to be a big fight and they could receive fees of millions of dollars, unless they agree to work on a contingent basis. If justice is achieved after so many years, when the horses have left the stable, it is not effective. The parties could achieve a settlement in the meantime, but then the plaintiffs will receive only a fraction of what they have lost. In fact, this time they have received an opponent, who will make them yearn for Gorekius and Istovius, who were not as audacious as Zrontius.

Zrontius has not the scruples of Gorekius and Istovius, he is a self-made man, a millionaire, who apparently has managed to achieve his business successes legally, maybe to the detriment of many parties, but legally. And really, if we analyze the legal arguments of this lawsuit, not the ethical ones which are blatant, they are very weak. Erinsar has never committed to buying the shares of Soktow. Many lawsuits were filed against Soktow, including the one of Astossg, which may endanger the cash of Soktow. The return on investment on the cash, about five percent, is minimal. Zrontius has stated from the start that he was going to invest the monies in his real estate business. Even Istovius in an interview said that you can earn a lot in real estate. If the investments that Soktow is going to finance are risky or not is a matter of opinion. After all, the investments of Soktow in R&D were also very risky.

Zrontius, Gorekius, and Istovius have not moved an inch without receiving the best legal advice. The selling of Istovius' shares of Erinsar to Soktow is brilliant, as it is at the expense of the minority shareholders of Soktow, it may cause a further collapse of the shares of Soktow, Zrontius may purchase even more shares of Soktow, but this time at less than $5, from the discouraged shareholders who will try to cut their losses. After all, why not, it could be the final chord to the symphony of Soktow, which was purchased by Erinsar ten years ago at exactly the same price before the reverse split. No, definitely, a lawsuit is not the right answer. In this case, as in all the other cases, the only answer is the ethical answer, the strike of the shareholders who will never buy anymore shares of Nalodo, Erinsar, Durtem, Ertel or companies that are managed by individuals like Gorekius, Istovius or Zrontius, all honorable men, but not so nice to their minority shareholders.

Could it be that the maxim 'Forget and move on', quoted in the article at the beginning of this case is no more the motto of the business world in Israel? Retaliation would seem to be a luxury you can afford, especially when you rally the largest organizations in Israel against the wrongdoers. Those bodies are no longer afraid of getting a reputation for vindictiveness, avenging lost honor and money may not be 'passe' after all. Istovius must have thought that 'the net present values, at any reasonable discount rate, must work against honoring obligations' toward Karisios who was 'insignificant', but when he and Gorekius did not honor their obligations to purchase the Soktow's shares at $14, the strongest Israeli companies sued him. Unfortunately, our tolerance for broken promises encourages risk taking, as Bhide and Stevenson state in their article, but this tolerance was permuted from risk taking in the dynamic entrepreneurial economy meant by the distinguished authors into risk taking in financial "combinatorics" to the detriment of the minority shareholders. The 'risk takers' have no more fear of prison and stigmas as in the past, but in view of the permutation of the leniency toward them one wonders if the time has not come to resort once again to the good old methods of deterrence as prisons and stigmas. These would deter marginally honest people from making the salta mortale, which could cause them the shame of prison and stigmas.

The sale of the shares in Erinsar by Istovius to Soktow is maybe the most flagrant example of the wrongdoing to the minority shareholders of Soktow and Erinsar. Istovius, we remember, has received 676,709 options to purchase Erinsar's shares at $7 per share in January 1999. A few weeks later, the controlling shares, about 37 percent of the company, were sold to Nalodo at $16.9. Subsequently, about 10 percent of Erinsar's shares were sold to Ertel at about $11 per share. The closing date on the agreement of the purchase of Erinsar's shares, signed on February 25th, was on May 4, 1999. In the June 1 press release of Erinsar on its quarterly financial statements, it was divulged that on March 25 an interested party, probably Istovius unless somebody else also received options for 676,709 shares..., exercised options for 676,709 shares. The benefit in the amount of $2.2 million ($3.25 per share) was reflected in the additional expenses of Erinsar for the first quarter of 1999. On April 13, Erinsar distributed a dividend in the amount of $2 per share, to which after reading the relevant documents, Istovius was probably entitled, thus giving him an additional benefit of $1.4M.

Finally, we know that Istovius sold all his Erinsar shares at $10 a share for a consideration of $8.8M. Those shares include, as we have seen, also the new 676,709 shares. The press releases do not clarify if the options were given at an exercise price of less than $7, or that the benefit on March 25 was a result of the difference between the exercise price and the market price on this day. However, it is obvious that the benefit of Istovius was at least $3M, as he sold his shares at $10 and probably received the dividends of $2. He did not sell his shares to Zrontius at $16.9 or even $11, as Zrontius had already gathered about 50 percent of the company's shares, but Istovius sold his shares to Soktow at the highest price in two months, and the minority shareholders of Soktow did not receive a satisfactory explanation as to why Soktow had to buy the Istovius shares in its parent company, what benefit was derived to the company and to them from this purchase and why did Istovius deserve a benefit of $3M within a few months when nobody believed that Istovius would continue to work in the companies Soktow and Erinsar. Istovius resigned on June 15 as a Director of Soktow. On July 1, 1999, Ertel appointed another person as Chairman of the Board of Soktow.

A partial answer to those questions can be received from the Israeli newspaper Yarmuk on July 26. We learn that Erinsar announced that Soktow had founded a new company Ibsol, in joint venture with Istovius. Soktow is purchasing Istovius' shares in Erinsar in three years for $8.8M. Out of the 3.8 percent of Erinsar's shares held by Istovius, 0.9 percent were already purchased by Soktow. Out of the consideration that Istovius will receive, $4.6M will be allocated to reimbursement of the loans taken by Istovius to purchase part of Erinsar's shares and $2.2M will be invested by him in the new company. Soktow has committed to invest in Ibsol a similar amount and allocate to the company shareholders' loans in additional amounts. Furthermore, Zrontius will give to Istovius a special bonus that will probably amount to $5M for saving Erinsar $50M out of the forecasted expenses on Erinsar's sale of its activities to Gosstik. We remember the very high amount that was forecasted for taxes and expenses on the Gosstik deal. This huge allocation has moderated the increase in Erinsar's share price and the amount of the dividends distributed to the shareholders. But, suddenly, after the company is sold to Zrontius, it appears that Erinsar can save $50M (!) from this allocation and a modest bonus of 10 percent is only a moderate reward to Istovius' dexterity.

Did Nalodo, Gorekius, Istovius, Ertel, Zrontius and the other interested parties know about this potential saving, which is a net profit, in their negotiations? Did the minority shareholders know about it? How did it affect the consideration to all those who sold their shares at $11 or $8? What could have been the impact of an increase of profit $50M in the Price to Earnings ratio of Erinsar, if it was done prior to the sale of Erinsar to Ertel? How did it happened so conveniently a year after the sale of the activities to Gosstik that this saving was achieved? It is probably legal to found a company of Istovius and Soktow, it is very legal for a subsidiary to buy the shares of its parent company, it is also legal to distribute a bonus of 10 percent of the savings to a brilliant negotiator, but why on earth those legal actions always work to the benefit of the interested parties, the affiliated shareholders, and to the detriment of the minority shareholders of Soktow and Erinsar?

Is it an axiom that they must always lose? Indeed it is, which brings us to the axiom of the unethical companies: In unethical companies, minority shareholders will always lose in the long run. The answer to this axiom is obvious – only ethics can work in favor of the minority shareholders. Ethics will give the answer if benefits of more than $8M to the CEO of Soktow and Erinsar are reasonable while the only ones who do not benefit from his activities are the minority shareholders. The activist associations will prevent such unethical acts and in the meantime minority shareholders are warned not to invest in unethical companies and to invest only in ethical funds or refrain altogether from investing in the stock exchange. This is the only way to break the Gordian knot that strangles the minority shareholders and to force companies like Loskron, Erinsar and Nalodo to be ethical.

A shareholder class action suit was filed on March 16th in the Supreme Court of the State of New York seeking to enjoin the proposed freeze-out of the minority 43 percent public shareholders ownership of Soktow at $14 per share by Erinsar, the 57 percent majority shareholder. The suit names as defendants Istovius, Gorekius, ..., Erinsar and Soktow, and seeks damages in the event the transaction is consummated. The complaint alleges grossly inadequate consideration for the 43 percent minority public shareholder interest, inadequate procedural protections for the public shareholders, and that defendants have engaged and are continuing to engage in acts of self-dealing, unfair dealing, gross overreaching and breaches of their fiduciary duties, all in an effort to enable the defendants to acquire the 43 percent minority public ownership for as little as possible. The lawyers firm that filed the class action was Sammel, which specializes in those matters.

Another lawsuit was filed in Israel against Erinsar and nine of its directors on April 14 alleging discrimination against the company's minority shareholders. The plaintiffs also proposed that the lawsuit be treated as a class action. The remedies sought by the plaintiffs include a distribution of additional cash dividends to shareholders or, alternatively, either redemption by the company of all shares held by the plaintiffs or a tender offer of the shares held by the public. In addition, the plaintiffs asked the court to restrain the company's directors from using the company's funds in any way not connected directly in the field of k. Erinsar stated that it could not at this stage assess the results of the lawsuit and its effects on the company, if any. The company secretary, Otto, said: 'The company believes that the claims raised in this suit are without merit, and we will vigorously oppose them.' This lawsuit was filed by the insurance companies, including a subsidiary of the Durtem Group, which did not succeed in selling their shares of Erinsar to Zrontius.

In the investment community in Israel, class actions are very rare. Since it was permitted to file class actions ten years ago, only a few dozen lawsuits were filed and none of them ended in a judgment. Most of the suits were dismissed immediately by the Court, some of them were prosecuted over many years in the Israeli courts, a few of them have ended in a compromise. These facts show that the Israeli court turns a 'cold shoulder' to those actions, and does not encourage investors to resort to this vehicle. In this context the many class actions that were filed against Soktow and Erinsar are a milestone, especially due to the important bodies that have filed the class actions. It was uncommon in the past to sue business colleagues, but this time they have joined forces to sue Nalodo, Erinsar, and Ertel. Until now the institutional investors used to come to shareholders assemblies to vote against wrongdoing and sell their shares or cease to invest in companies and groups that have harmed them. Is this move a step toward safeguarding the interests of the minority shareholders? It is too early to tell, but the writer of this book is very skeptical that it will contribute to it, in view of the attitude of the Court, past experience, and the length of the deliberations. The Internet was also not an adequate vehicle for safeguarding the interests of the shareholders in this case, as the response on the thread of Erinsar and Soktow, both of them traded in New York was very poor, without any proportion to the response of the shareholders of Mastoss, that we are about to see in the next case.

Astossg filed a lawsuit on April 15 against Soktow. "Astossg, the former CEO of Soktow, who found himself in direct conflict with Istovius, the CEO of Erinsar and Chairman of Soktow, has left the company angrily... Astossg is suing at the Haifa Labor Court Soktow and its directors for 10 million shekels... Astossg states that he had to leave the company in 1998 as a result of an unlawful and immoral move. He maintains that after leaving it was promised to him that the company would know how to appreciate his contribution and compensate him in consideration of the high price and damage that he has suffered from his leaving. Instead, he received a shameful proposal, that did not stand in any proportion to his contribution in the 27 years of work in Soktow. Astossg promises to disclose in the trial unlawful acts that were done behind his back, and that enabled Istovius to achieve his leaving. Astossg says that his leaving made possible the dismantling of Soktow and sale of parts of it in order to fill the parent companies Erinsar and Nalodo with cash, and in order to achieve personal gains. Astossg means the bonus that Istovius received in the amount of 16 million shekels, after selling the assets of Soktow and Erinsar. 'As I have known in the past to invest all my efforts in contributing to Soktow, thus I will focus now in disclosing the evil and greed that have brought to the dismantling of Soktow, while trampling my rights and tarnishing my honor', declared Astossg yesterday." (Yarmuk, 16.4.99) This lawsuit follows the alleged dismissal of Astossg, as reported by the newspapers of September 2, 1998 that wrote that Astossg was dismissed by the Board of Directors of Soktow on the 1st, and that Istovius replaced him as CEO of Soktow. Astossg replied a day later that he was not dismissed and that he was forced to leave the company. In the press release of the company it was stated that Astossg 'has concluded his duties as President and CEO.' Vive la petite difference.

This case deals with ethics. Ethics stem mostly from people, from the character of the executives running the companies. We have described the motives of Astossg, the Soktow's CEO and hero of this case, who left Soktow in September 1998, as he did not agree with Istovius and Gorekius' plan to dismantle the company. The Board of Directors supported the affiliated shareholders against Astossg, it is unknown if any one of them, including the outside directors thought about the minority shareholders. Astossg was interviewed in Yarmuk on September 15, 1998, after the decision was taken to sell the activities of the company to Gosstik and Priam: "In the middle of the night, after the Board meeting where it was decided that Astossg will terminate his duties as CEO of Soktow, Astossg drove his Rover car to his house in ... Haifa, and thought about his predecessor Poftrim. Poftrim, remembered Astossg had suffered more than four years ago from a head on confrontation with Istovius and when he came to the conclusion that there is no hope for the company he slammed the door and left. 'He said that Istovius wants to dismantle Soktow and sell it and that will be the end of the company. We did not believe him, we thought that he was delirious, that he had a personal agenda against Istovius. Poftrim was very angry, he ceased our relationship although we were close friends and our kids grew up together. He was my first boss. He hired me and promoted me, but since he left we didn't exchange a word. Today I know that he was right. If I would meet him now, I would ask for his forgiveness for not believing him.'"

On the other hand is Gorekius, who was interviewed in Yarmuk on August 8, 1999: "It was hard for me to write the retirement announcement', said Gorekius retiring from the CEO's position of Nalodo... I cannot see myself not functioning and contributing something to the society I am in. When I say contributing, it may be not always in a positive way, as I have done also bad things, of course unintentionally... Within a few years Nalodo has grown and ramified in subsidiaries that made a breakthrough of the Israeli high-tech industry in the world. Erinsar, Furolias, Soktow are only part of the companies that employ thousands of engineers and programmers and sell more than a billion dollars annually... The outstanding performance of Nalodo has won the founder and president of Nalodo the titles of 'King Gorekius', 'Mr. High-Tech', 'Henry Ford of the Israeli High-Tech', given by the Israeli media throughout the years. It seems that those titles do not intimidate Gorekius. On the contrary, it seems that he thinks that they were tailored to his size. 'It made me feel good, why not, but I have learned to be indifferent to those things.'... I have never led the list of the best-paid executives in Israel, I was always in the fiftieth or sixtieth place. I have no doubt that I could have won a lot more money, but I think that it is a matter of character. It is not that I did not fight for it enough, but that it was not so important to me. I am sure that it is possible to analyze my conduct toward remuneration as a weakness... The company had many ups and downs, obviously, and the shareholders punished me, what do you think, the shares went down. But the stock market is not relevant in the short run. Only in the long run."

What could Gorekius say, in the long run, to all those shareholders of Soktow, who bought their shares at prices of $20, $25 or more and had to sell them in the long run at about $1 after its assets were sold; to the shareholders of Furolias who bought their shares at a valuation of $100M and had to sell it to Erinsar in the long run at a valuation of $8M; to the shareholders of Erinsar who lost $100M in valuation recently after Nalodo sold the controlling shares to a real estate contractor? Yet he is called Mr. High-Tech and not a financial manipulator and the minority shareholders are called speculators. In the long run who won more? Gorekius personally who ranked 'only' on the fiftieth place but who received the equivalent of ten million dollars as a retirement parachute, besides the share valuation of about $15M that he held at his retirement, on top of the millions he has sold throughout the years and his exorbitant profits on Memnit shares; Istovius personally with a better ranking and similar capital gains; the Durtem Group with hundred of millions of dollars in capital gains on Nalodo's group shares, or the despicable minority shareholders speculators, many of whom lost almost all their investment? What is the long run, when will be the end of double talks? How this conduct compares to the conduct of Karisios, Poftrim and Astossg who did not sell in the long run most of their shares in Furolias, Soktow and Erinsar, as they believed in the companies and did not want to use insider information. Who should really be Mr. High-Tech, people like Astossg and Poftrim who managed for many years the flagship of the high-tech industry and who had an impeccable ethical reputation, or the GIs?

Astossg filed on September 1 1999, exactly a year after leaving his position as CEO of Soktow, a $100M class action in the Tel-Aviv district court against Soktow, Erinsar, Nalodo, Istovius, Gorekius, B,... Astossg held 0.3 percent of Soktow's shares. Gorekius, curiously enough, did not have any shares in Soktow according to the press. So, 'run and sell' your shares according to Istovius' law, lasciate ogni speranza ye minority shareholders who remain with shares of a company managed by the GIs, in which they do not have any shares. In the Yarmuk article 'Astossg states that the directors of Soktow, and especially its Chairman Istovius, have managed Soktow in a discriminatory way toward its minority shareholders, and acted for the benefit of Erinsar in order to divert Soktow's funds to it and to affiliated companies to the detriment of Soktow and its minority shareholders. This, in transgression to their fiduciary duty and thus hurting Soktow's business, assets, equity and valuation. Astossg states that in '96, Erinsar bought Soktow's ... division for a price of $8.1M, although the valuation made by ... was $18-20M. In early '98 Erinsar sold this division to Gosstik for $20M.' (This is one more example of the miraculous exploits of Istovius, who buys companies and sell them a few years later for three times more. He did it also with Furolias sold to Mastoss, and with Deon sold to Gosstik, only that the minority shareholders of Furolias, Soktow, and partially Erinsar did not benefit from these exploits). 'Astossg states that the consideration received by Soktow in

November 1998 for the sale of the ... divisions to Gosstik for $100M and ... division to Priam for $269M is totally unreasonable and extremely low. The considerations received by Soktow and Erinsar, was divided according to Astossg between Erinsar and Soktow in an artificial way that does not reflect the intrinsic value of the assets, in order to pay Soktow a lower amount than what was due to it and divert to Erinsar a higher amount than what was due according to the intrinsic value of the assets sold. Astossg states that following his objections to the sale of Soktow's assets, Soktow's Board of Directors has decided to relieve him of his functions. Istovius' actions were done according to Astossg with 'fraud, conspiracy and conflicting interests'. Astossg requests the court to force Soktow and the other defendants to buy from him and from the other shareholders of Soktow their shares at a price of not less than $28.'

This article was written in the first economic page of Yarmuk on September 2, and had many repercussions. The photos of Astossg and Istovius appeared side by side, the public was impressed by the high amount of the class action - $100M, Astossg's allegations and former position as Soktow's CEO, and the fact that Astossg asked for a valuation that was about four times more than the market valuation of the company. He was not contented with Erinsar's valuation for the minority shareholders of Soktow - $14, but asked for a double valuation - $28, among others because of the alleged diversion, of the considerations of Deon and the Soktow's assets sold to Gosstik, between Erinsar and Soktow; events that Poftrim and Astossg have alleged prior to then in the Israeli press. But the most striking note is that Astossg filed the class action alone. No other economic body or individual joined him. Astossg found himself exactly in the same situation as Karisios, four years ago, that wanted to sue Furolias and Nalodo for $40M, but nobody wanted to join him. Karisios decided to back off, as he thought that he had no chance and. no funds to pursue this lawsuit all by himself, Astossg dared to go on and filed the lawsuit. He benefited from front page headlines for a few days and all the ensuing trouble of a lawsuit subsequently. What is the right approach for the minority shareholders? Will Astossg win the case within a few years in spite of the 'cold shoulder' of the Israeli courts? Does he have any chance against the batteries of the best lawyers in Israel that Nalodo and Erinsar have? Will he have the funds, health and peace of mind to conduct his courageous fight? Will he be perceived as a Don Quixote, a whistle-blower, or a true hero?

Astossg received an immediate 'interim' answer within a few days. We remember how the perplexed Karisios was afraid of suing alone Furolias, as he feared of being sued himself, as the group has done in the past. Astossg was sued within a week by Dargokks, who held 0.164 percent of Soktow's shares. This gentleman filed, according to Yarmuk of 9.9.99 (definitely a dangerous timing...) in the Haifa district court a class action of $148.4M (...) against Soktow, Erinsar, Nalodo, and their directors – Istovius, Gorekius,

Bsosskins, …. and Astossg (!) Dargokks stated that by the end of August `99, he spoke to Astossg and wanted to discuss various data that he had gathered toward filing a class action. This, in view of his concern that Soktow has done unlawful actions. Astossg, says Dargokks, tried to dissuade him from filing the lawsuit and wanted to verify with him various data. They decided to meet and go over the data. To his surprise, Dargokks was astonished to see that Astossg has managed to file the lawsuit ahead of him in order to protect himself. Astossg, who was a director in Soktow, took part in the decisions that damaged the shareholders. Dargokks follows most of the allegations of Astossg, focuses on the diversion of the consideration of Gosstik from Soktow's assets to Deon's assets in order to defraud the minority shareholders of Soktow, and adds to it that Erinsar has decided to back off from its offer to purchase Erinsar's shares at $14 on September 7.

Who is this Dargokks, and how comes that he decided to add Astossg to his class action that is 48 percent higher, only a week after Astossg filed his lawsuit? To accuse Astossg of collaboration with Erinsar, when he opposed the sale of Soktow's assets so fiercely, on the 'formalistic' ground that he remained for a short while director of the company after he left it, is very, very far fetched to say the least. What does it matter who files first the class action, and how does Astossg protect himself if he files first his lawsuit? Can he be accused of 'lack of good faith' because of it? How can we explain the perfect timing of Erinsar's withdrawal of the 'alleged' offer immediately after Astossg's class action, on September 7, implying that this class action and others jeopardize the attractiveness of Soktow and put a high risk on the acquisition of Soktow's shares. This is risky, and the purchase of assets from Ertel with a multiple of 1,600 a few days later, on September 9 is not risky. This sequence of events permits Zrontius to save $100M he should have paid to the minority shareholders of Soktow, lessens the risk of paying the shareholders if they succeed in their class actions, as Soktow and Erinsar will remain without cash after acquiring Ertel's assets at exorbitant prices, as we have seen in the details of the lawsuit of November 1999. Definitely, there are so many lawsuits that one can get confused.

Within a week from Astossg's class action, he is sued himself for a higher amount than he has sued and in addition to the costs and management attention on his lawsuit, he has to defend himself from a 'fellow shareholder'. And in the midst of this multitude of class actions that have been filed and were filed subsequently the only ones who probably were not worried and grinned at the divided forces and the weak legal arguments were Nalodo's and Erinsar's executives, who were confident that their lawyers will defend them perfectly. And why not, the best lawyers of Israel have also succeeded a few years ago to prevent the bank managers who cost the state of Israel losses of billions of dollars from going to jail even for the minimal period of six months that they were sentenced to. The heads of the Durtem Group were spared from the jail sentences and probably were confident that their subsidiaries' managers would not encounter any risk whatsoever.

The articles in the newspapers became less and less polite toward Erinsar and Nalodo. In an article in the Israeli newspaper Horton on September 9, describing the lawsuit of Dargokks and the withdrawal of the offer of Erinsar to purchase Soktow's shares, we read: "The allegations about the conduct of Nalodo toward the shareholders of Soktow are heard for years. Nalodo tries to explain all the time that its conduct is impeccable, but the fact that the two last CEOs of Soktow, Astossg and Poftrim, have left the company with allegations that the controlling shareholder of Soktow plunders its assets and damages it, throws a heavy shadow on the conduct of Nalodo and Erinsar toward the minority shareholders of Soktow. Erinsar is a public company that has to care for its shareholders, but as the controlling shareholder of Soktow it has a fiduciary duty also for the minority shareholders of Soktow, and it is very doubtful if it has fulfilled this duty over the years." To the practical shareholder, it should have sufficed that Gorekius also did not have any Soktow' shares, according to the press, as he had only a very small amount of shares of Furolias, which he received many years before the takeover by Erinsar, at an even lower price than the outrageously low price of Erinsar's offer to the minority shareholders of Furolias. But of course, Gorekius stated in one of his latest articles that he did not care much for money, although it was not clear if he meant his money or the minority shareholders' money.

Zrontius was more blunt, referring to the complaints of the shareholders of Soktow, whose company bought real estate from Zrontius for $196M and of Erinsar, whose company was in negotiations to buy more real estate from Zrontius, he said in Horton on September 10: "The shareholders who do not like the deals can sell their shares." And Rorton, the former lawyer of Zrontius and the present Chairman of Soktow, commented that "the decision not to carry on the offer to purchase the shares of the minority shareholders of Soktow, was taken by the independent directors of Soktow, and that he and Zrontius had nothing to do with it. Furthermore, in the last few months they worked hard to find what were the best investments for Soktow's cash surplus of hundreds of millions of dollars. And finally they found it – investing into the real estate business of Zrontius.' It reminds us so much of what happened with Furolias, that we could devise a new law, the GZ laws which consists of:

- Finding independent directors who will decide that what is best for the minority shareholders is to do what the affiliates want to do – merge Furolias with Erinsar at ridiculous prices, or not buy Soktow's shares by Erinsar at reasonable prices.
- In parallel, working very hard for months to find investors for Furolias (Wersnon mentioned to Karisios - 15 to 17 companies) until they find the

sister company Erinsar or to find adequate investments for Soktow until they find the investments of the parent company Ertel.

In both cases, on sauve les apparences, we keep everything legal and unethical, as the decisions are taken by the independent directors, and the CEOs painstakingly review all the alternatives. The SEC does not investigate, the auditors do not interfere, but, surprise, the new Company Law in Israel makes a fantastic step forward by changing the name of the 'outside' directors to 'independent' directors, the press writes some sarcastic articles, but the shareholders have only recourse in expensive class actions or in an ethical revolution.

And Erinsar announces to Soktow (like Gorekius announced to himself in Furolias' case where he held the three positions of the parties involved as Chairman of Furolias, Erinsar, and Nalodo) on September 7: 'Following extensive deliberations and contacts with the Committee of Independent Directors of Soktow, the Committee of Independent Directors of Erinsar, which had been appointed to review and consider the proposed transaction, resolved that Soktow's financial condition and financial exposure resulting, mainly, of the contingent and other liabilities relating to the sale of Soktow's assets in 1998, had led them to conclude that the proposed transaction would not be in the best interests of the shareholders of Erinsar.'

Erinsar states also on the same day, referring to Astossg's class action that 'the Company is of the opinion that the Claim is groundless, is devoid of both factual and legal merit, and intends to vigorously oppose and defend same.' On September 22, Erinsar refers in its press release on Dargokks's claim stating: 'It shall be noted that the subject matter of the Claim is similar to a different claim, submitted by Astossg, former President of Soktow. Astossg himself is one of the defendants in the Claim. In the Claim the Plaintiff challenges, inter alia, Astossg's authority to submit the mention previous claim. The Claim is groundless, etc, etc, etc.' On October 19, Erinsar announces that it has purchased in the last six months 1.6 percent of Soktow's shares, 272,400 shares, at a consideration of $2.5M. In this period, the price of Soktow shares has fallen from $11 to $7, on the average - $9.2. So, why buy shares at $14 or $28, as stated in the class actions, if you can purchase them at $9 or even $7 in the stock exchange.

Erinsar is not deterred, Ertel is not deterred, but Kamtov, the largest company in Israel is deterred from its audacious move to sue their colleagues at Nalodo, Erinsar, and Soktow, the subsidiaries of its colleagues in the Durtem Group. We remember the very favorable response of the press on the class action that the Israeli institutions filed against the Nalodo group on November 3, 1999 in order to force Erinsar to fulfill its 'promise' to buy Soktow shares at $14. The congratulations were apparently premature, as Kamtov decides a few days later to back off. It reminds one so much how all the important companies backed off in Furolias' case, that one could devise another law "the back off law", stating that there is always a way to cause important bodies to back off from suing strong companies for wrongdoing against minority shareholders.

The companies could back off as a result of a multitude of reasons such as:
intimidation, everybody lives in a glass building,
you sue me today I will sue you tomorrow as nobody is perfect,
if you back off I will compensate you in an 'unrelated' deal,
it is not done between old buddies,
we are partners in so many businesses so how can you sue me,
believe me it is not suitable that you will converge with this bunch of speculators,
noblesse oblige, if the 50 richest families in Israel will not stick together the mob will rule the state,
it is a waste of time, who can afford it, lose and move on,
I give you my word of honor that everything is legal,
I will fight this lawsuit with all my power,
I have just retired and you are spoiling my peace of mind,
I did not know of the wrongdoing but I have to back up my employees,
etc.

'We can no more write our article. For four days we are intending to write this article, four days that we are thinking of the greatness of the moment, four days that we are pondering on the change that is happening in the Israeli economy. And then, we can no more write our article. Last Tuesday we were astonished from the class action of the largest institutions against the Erinsar – Soktow group, Zrontius, and the Nalodo group that is controlled by the Durtem Group. We have seen many class actions, we have filled in many newspapers – but such one we have never seen. This time it was not..., a professional troublemaker that files many class actions, not even ... who dared once to go against ..., and not also ..., a driving teacher from ... that went against .... For the first time the plaintiffs that sue Zrontius are the top of the Israeli stock exchange: The pension funds of Bank Horshrem, the brokers ..., Kamtov, and many more funds and institutions...

The second part is twice as much surprising: The plaintiffs have stated that Nalodo and its directors have infringed their fiduciary duty toward the shareholders when they sold their shares in Erinsar, kept the cash pot in the companies, and have chosen not to distribute the money to the shareholders, although it was not used for adequate purposes. That is why we wanted to write that this class suit is a turning point in the Israeli market: For the first time institutions of the first league go against what they perceive as the plunder of the money of their clients and colleagues. We wanted to write that for the first time they treat this matter seriously: They join forces and hire a first class lawyer Naran. We wanted to write that for the first time a class action is filed by serious plaintiffs who are not afraid to attack the defendants in harsh words, although they are amongst the most famous businessmen in Israel.

And most of all we wanted to say 'chapeau!', well done!, to … the President of Kamtov, because in the plaintiffs we found surprisingly Kamtov for a small amount of shares that they have in the Erinsar group. We wanted to say that it indicates the new norms of the market: Kamtov owned by the family … joins a class action against the Durtem Group owned by the family …. We wanted to say to … that although he and the … meet three times a month in cocktail parties, in spite of the smallness of the Israeli market, he makes a distinction. We wanted to say that maybe the days of 'I'll not sue you and you'll not sue me' are over. But the day before yesterday the stock exchange received a short notice: It was only a technical error, explains Kamtov. 'Kamtov removes itself from the list of the plaintiffs in the class action, and as far as it is concerned it removes Kamtov's lawsuit….'
We can no more write our article. (Horton, 9.11.99)

Noblesse oblige has won, the aristocrats are not going to join forces with the untouchables. The list of the deserters will probably increase in the future and ultimately Astossg will remain alone, with Dargokks to deter him, but who is this Astossg anyway?

The heroes of these cases are lonesome riders, whether it is Mme. Neuville of l'ADAM struggling against huge opponents of Loskron, or Karisios from Erinsar and Furolias struggling against his old friends who have betrayed their ethical values, or Poftrim and Astossg of Soktow who managed the company but could not accept the norms of Gorekius from Nalodo and Istovius from Erinsar, or finally Mr. Pink, the virtual hero of the Internet, who discovered alone in the whole thread of the Internet the wrongdoing of Mastoss. All of them fought for the minority shareholders, while keeping their impeccable integrity and ethical norms. Nobody believed them when they spoke, yet if something would have been done, it would have prevented more acute wrongdoing. All of them were alone in their struggles, all of them suffered extremely personally or financially for their beliefs, but the most tragic hero is undoubtedly Astossg, as he brought his struggle to the extreme. Tirelessly, in the newspapers, in hundred of conversations, in lawsuits, he continued his struggle, he pursued his ordeal. He was not deterred by his mighty opponents, by the financial hardships of the trials, by the tarnished reputation of a whistle-blower or a Don Quixote. Those heroes, each and every one in his modest way, have started to make a change, a small step toward achieving the goal of redressing the wrongdoing to the most forgotten public, the minority shareholders.

EPILOGUE OF THE CASE

The author of this book on ethics had an ethical dilemma. On the last day of editing the manuscript, just before sending it to the publisher, he saw a small article in Yarmuk on November 20, 2000. It was so small, only a few lines, that it probably went unnoticed by the readers. The article stated that Astossg had withdrawn his two lawsuits against Soktow and its directors after reaching an agreement with the company. We remember that Astossg filed two lawsuits, one at the Haifa Labor Court on April 15, 1999, for 10 million shekels (about 2.5 million dollars) stating that he had to leave the company in 1998 as a result of an unlawful and immoral move, and reminding the court of the bonuses of millions of dollars that Istovius received for the sale of Soktow's assets. The other lawsuit was a class action of 100 million dollars. As court decisions are made public in Israel, a short inquiry revealed the verdict of the Labor Court, which validated the agreement between the parties. Soktow agreed to pay Astossg the sum of $970,000 before taxes as a severance grant, without acknowledging any of Astossg's accusations, and the parties agreed that they would not have any mutual claims. Astossg compromised for a net after tax amount of less than half a million dollars, which is quite low for a normal grant for a CEO quitting a large company. But Soktow generously agreed to leave him his Rover 827 company car...

We remember the shares, bonuses and severance payments of tens of millions of dollars that the other protagonists of this drama had received. But they cooperated with the owners, while Astossg rebelled. As the class action was withdrawn, no compensation was received in this lawsuit by Astossg or any other minority shareholder for the wrongdoing that was committed by the affiliated shareholders. The author of the book therefore had an ethical dilemma; should he disregard this last-minute news and leave readers with the hope that Astossg might win his case and the minority shareholders might recoup part of their losses? Or should he be consistent with full transparency and disclose all the facts? In a final-hour epilogue, he chose to disclose the outcome of the case, which proves that class actions, the press and transparency are not yet adequate vehicles to safeguard the interests of minority shareholders, at least not in Israel 2000.

But the lonely rider fight of Astossg made an important impact. He disclosed publicly for the first time the major wrongdoing of one of the mightiest business groups in Israel. He suffered from his sincerity, but the minority shareholders received a fair warning not to invest anymore in this Group's companies. He paid a very high price for his integrity, honesty and loyalty to the stakeholders of Soktow. But Astossg brought business ethics to the forefront of Israel's collective consciousness and in that respect he was much more successful than Karisios. His Odyssey was therefore not in vain!

11

CASE STUDY OF THE AMERICAN COMPANY MASTOSS

"He who walks righteously and speaks what is right,
who rejects gain from extortion and keeps his hand from accepting bribes,
who stops his ears against plots of murder
and shuts his eyes against contemplating evil –
this is the man who will dwell on the heights,
whose refuge will be the mountain fortress.
His bread will be supplied, and water will not fail him.
Your eyes will see the king in his beauty
and view a land that stretches afar.
In your thoughts you will ponder the former terror:
'Where is that chief officer?
Where is the one who took the revenue?
Where is the officer in charge of the towers?'
You will see those arrogant people no more,
those people of an obscure speech,
with their strange, incomprehensible tongue."
(The Bible, Isaiah, 34:15)

Apparently much has not changed since the time of Isaiah, as the men who dwell on the heights are not necessarily the people who walk righteously and reject gain from extortion. The righteous people normally live in the plain in cozy three bedroom apartments and have a high mortgage. But Isaiah must have prophesied the plots of the stock exchange scams, asking the relevant questions of where is the chief (executive) officer, where does he hide from the angry minority shareholders, after he took all the revenues of the company to himself and to his affiliated colleagues. Those arrogant officers are seen all over the place, talking high and doing low, speaking in an obscure speech that can be understood in different ways to facilitate their schemes, with their strange incomprehensible tongue, the financial data dialect that nobody but them understands and enables them to wrong the shareholders legally with the blessing of the auditors, the assistance of the lawyers and the closed eyes of the SEC. This book is trying to make a change in order that our eyes and not the eyes of our great grandchildren will see the king in his beauty and view a land that stretches afar, the king of justice and the land of ethics, in France, the Netherlands, the United States, Great Britain, Australia and Israel, all over the world, the promised land in God's Kingdom!

This case will be based primarily on information retrieved from the Internet. While in 1994 - 1995, when most of the events of the first two cases, Loskron and Furolias, took place, the Internet was not currently used for analyzing companies and holding e-mail talks between shareholders. The case of Erinsar and Soktow took place mainly in 1998 and 1999, when the Internet was already in use for such purposes. However, in Erinsar and Soktow's case, maybe because it dealt with Israeli companies, the use of the Internet was very restrained and it did not reveal very important data. The Mastoss case, which takes place mainly in 1998, proves the importance of the Internet as a vehicle for spreading vital data to everyone, although it contains also a lot of misinformation that the reader is invited to discern out of the genuine information.

Mastoss is a US high-tech company located in California. In the 1996 annual report of Mastoss (SEC form 10-K) it is disclosed that in 1995 the company augmented its activities with the acquisitions of certain assets of two Israeli companies which resulted in charges of $6,211,000 and $1,465,000 for purchased technology in progress and restructuring. In May 1996, the company acquired 50 percent of the outstanding stock of a company located in Italy and in September 1996, Mastoss completed the Furolias acquisition, acquiring assets related to Furolias' activities in Germany, the US, the UK, the Netherlands and Israel.

As stated in Loskron's case: la boucle est bouclee, all the cases tie up in the same coherent plot. Loskron is not connected directly with the others but the sequence of events is so strikingly similar to that of Furolias, happening also at the same time and using the same arguments for obstructing the rights for minority shareholders that the Loskron case belongs de facto to the trilogy of Furolias, Erinsar and Mastoss. The relation between Furolias and Erinsar is obvious, they have the same parent company Nalodo, and Istovius and Gorekius were encouraged to fulfill the Erinsar and Soktow case only after succeeding to implement all their plans about the purchase of Furolias and committing the wrongdoing to the minority shareholders without encountering any substantial resistance. Mastoss purchased Furolias, its founders came into close contacts with Istovius and it has possibly permeated into their corporate governance and their conduct toward their minority shareholders, as will be shown in this case.

Unfortunately, the lack of ethics is like a contagious illness. Once you start to work with unethical executives and companies, it influences your way of thinking and in many cases it causes you to lose your ethics as well. You see that crime does pay, that nothing happens to the unethical executives, that it is very convenient for surviving, that 'everybody' does it anyhow, and you start to behave unethically even if you are very ethical at the start.

The Furolias assets acquired included Furolias' technology in progress and existing technology, its marketing channels, its … products and other rights. This acquisition also resulted in charges in the amount of $17,795,000 and $6,974,000 for purchased technology in progress and restructuring, respectively. In September 1996, Mastoss completed a private placement of an aggregate of $30M principal amount of 5 percent convertible subordinated debentures due August 6, 1999. Proceeds from this private placement were used to purchase the Furolias business. It sounds familiar, this is exactly the same method that Istovius used two years ago in 1994 to finance the purchase of Furolias, but with one difference – Istovius purchased the whole company for about $8M (in shares of Erinsar) and he sold now only some of Furolias' assets to Mastoss for above $22M, retaining part of Furolias' technology for Erinsar. Subsequently, as a result of the SEC's position, Mastoss added a non-recurring non-cash charge to its results of operations for 1996 related to the issuance of the Debentures in the amount of $4,357,000.

Mastoss' 1996 revenues amounted to $89M compared to $39M in 1995. Net loss increased from a loss of $1.2M in 1995 to a loss of $9.7M in 1996, due to the Furolias acquisition and the restructuring costs and purchased technology ensuing. The auditor of Mastoss, Furolias, Nalodo and Erinsar is the same – Ascorage. Ascorage finished its report on Mastoss with the following standard sentence: "In our opinion, the financial statements referred to above present fairly, in all material respects, the financial position of Mastoss and subsidiaries as of December 31, 1995 and 1996, and the results of their operations and their cash flows for each of the three years in the period ended December 31, 1996 in conformity with generally accepted accounting principles." The assets of the company by the end of 1996 amounted to $97M, mainly current assets, its current liabilities amounted to $25M, long-term liabilities amounted to $19M, mainly convertible debentures, and Net Equity amounted to $52M.

The fully diluted amount of common shares was 18,377,000. The price of the shares in December 1996 was on the average $23, thus giving Mastoss a valuation of $423M, not bad for a company with an equity of $52M, sales of $89M and a net loss of $10M. But this is quite common in the US stock exchange at the end of the millennium for growth company as Mastoss, that doubled its operations every year mainly through acquisitions. The price of the shares even increased to about $33 in July 1997. Mastoss' valuation was three times higher than Erinsar (k.) or Soktow valuation, although they had sales of hundreds of millions of dollars and were profitable while Mastoss was losing money. But the financial community does not penalize the valuation of companies on account of technology acquisition and restructuring costs, that are extraordinary expenses, if the current operations are profitable. What happens in a case like in Mastoss' case where those extraordinary expenses are incurred every year and some of the shareholders believe that they are inflated on purpose in order to conceal current losses, is the subject of this case, as it bears a tremendous influence on the shares' price and the profits or losses of the minority shareholders.

We learn from Mastoss' prospectus of 9.9.97 that Mortishko one of the two founders of the company held 2,033,930 shares of Mastoss, 8.6 percent of the common stock owned prior to the offering, of which he sold at the offering 200,000 shares. Rostronsky, the other founder, held 1,944,811 shares, 8.3 percent, and sold 200,000, and Nartokow, the CEO of Mastoss held 968,437 shares and sold 100,000. All executives and directors as a group, 8 persons, held 5,444,652 shares prior to the offering, 23.3 percent, and sold 550,000 shares at the offering, about 10 percent of their holdings. The annual salaries of the three key officers were quite low in US terms, $100-110K each. They were also entitled to a bonus.

Among the risks divulged in the prospectus we learn of the risks associated with the recent acquisition of Furolias and potential future acquisitions. Furolias' sales in 1994 were $33.4M with losses of $11.6M (Erinsar probably took also maximum reserves in the year of acquisition as they are counted as extraordinary expenses), in 1995 sales increased to $35M generating a small profit of $79K, while sales in 1996 through September 25, the date of acquisition by Mastoss, were $19.5M with losses of $6.1M. Mastoss stated that: 'Future acquisitions by the Company could result in charges similar to those incurred in connection with the Furolias acquisition, potentially dilutive issuance of equity securities, the incurrence of debt and contingent liabilities and amortization expenses related to goodwill and other intangible assets any of which could materially adversely affect the Company's business, financial condition and results of operations and/or the price of the Company's Common Stock. Acquisitions entail numerous risks, including the assimilation of the acquired operations, technologies and products, diversion of management's attention to other business concerns, risks of entering markets in which the Company has no or limited prior experience and potential loss of key employees of acquired organizations. Prior to the Furolias acquisition, management had only limited experience in assimilating acquired organizations. There can be no assurance as to the ability of the Company to successfully integrate the products, technologies or personnel of any business that might be acquired in the future and the failure of the Company to do so could have a material adverse effect on the Company's business, financial conditions and results of operations.' (Mastoss' 9.9.97 prospectus, page 10)

This brilliant dissertation on the risks of future acquisitions hidden on page 10 of a bulky prospectus gave Mastoss full legitimacy to perform what will be divulged in this case without leaving the minority shareholders any legal recourse. Probably no investor read this warning at page 10 that is so relevant to what is about to come and they bought the shares of Mastoss hoping that the management of the company will be able to cope with the acquisitions that they made. The problem was that every year they made new acquisitions, which were larger and larger and the management took more and more reserves on their account, that even surpassed the amount of the acquisitions. When the acquisitions stopped, the company incurred operational losses and the shares' price dropped to $5. But by then, the company has already made an offering at a valuation based on a price of $35, the founders and executives sold a substantial amount of their shares at this high valuation and those who incurred the drop in the price were as usual the minority shareholders.

What have we in Mastoss' statement in the prospectus? The management foresees future acquisitions yet admits that it has limited experience in assimilating acquired organizations. The SEC would have compelled Mastoss to put in the first page of the prospectus a warning about a lawsuit of $20M, yet does not require the company to put in the first page a warning to its shareholders about its inexperience in handling acquisitions of new companies. The future will prove that from all the warnings this was the more relevant, yet it hid on page 10, although the company had an outstanding record of acquisitions and incurring extraordinary losses as a result of them. It is beyond the scope of this book to comment on the relevance of the SEC directives, yet this example of a substantial warning, that was not emphasized by the SEC, is a striking argument in favor of implementing ethics criteria in public offerings, as such a statement would not have met the ethics requirements of the activist associations, although it was completely legal.

On September 19 1997 Mastoss sold 2.9M shares of its common stock at a price of $35.75 per share. In total the market bought shares for $103,675,000 at almost the highest valuation ever of Mastoss' shares. Of the 2.9M shares bought, 2.35M were sold by the company (in the prospectus only 2M were planned) for $84M, and 550,000 shares were sold by the management of the company for $20M. The company has granted the underwriters a one month option to purchase up to an additional 435,000 shares to cover over-allotment. The offering was lead by top underwriters, such as Bonnty, one of the largest investment bankers of Wall Street. Mastoss increased the number of its shares outstanding to 25.9M (market valuation of $926M, almost a billion!). The founders and management of the company remained with less than 5M shares, less than 20 percent of the equity, thus making the unaffiliated stockholders the absolute majority with more than 80 percent of the equity. How they took advantage of their majority and strength is another question as will be analyzed later on.

1997 revenues amounted to $165M compared to $89M in 1996, with net income of $23M compared to a loss of $10M. Those were outstanding figures, the company continued to double its sales every year, incurred a very high profit of $23M, 14 percent of revenues, and had a price to earnings ratio of about 40, which is not extraordinary in Mastoss' product lines. This was the first time since Mastoss grew to be a large company that it had a 'genuine' profitability of $23M, without reporting extraordinary losses due to acquisitions. The operating income was even higher - $30M, as Mastoss took a provision of $9.5M for income taxes. Yet, a day after the publication of the 1997 results, on 24.2.98, Dan Spillane (most if not all the names of correspondents in the stock talks are fictitious) wrote the following opinion: 'I'm sorry to say, Mastoss is a loser. Let me explain... Mastoss is unfortunately not a special large-cap stock – so it is worth virtually nothing at all, given what would appear to be attractive financial ratios. And, given any little crack (loss of ... customers) Mastoss is worth less than virtually nothing. So there are really two stock markets. Don't be fooled – you're either 'worth infinity' or 'worth virtually nothing' in this market.' As prices were traded on the same day at about $25, $10 less than a few months ago at the offering although Mastoss achieved brilliant 1997 results, the stockholders were disappointed. Bold Man from Mars wrote on the same day to Spillane: 'Anyone please tell me what the heck happened to my beloved Mastoss??? I feel like Iraq being bombed by the US, it feels shitty...'

We can understand the reason of the drop in shares' price only after reading the comments written on the same day by Larry J. He reports that analysts explain the sharp drop of $5 in one day, February 24, 1998, to $23.9375, 17 percent in one day, as a reaction of the investors to news that Mastoss' inventories and receivables had doubled from year-earlier levels. Those are typically red-flag warnings about earnings, analyst .... said. For Mastoss posting its 32nd consecutive quarter of earnings and revenue growth, the inventory and large days sales outstanding escalation caught investors by surprise. Mastoss' CFO said 'he believes the company will meet analysts' estimates for 1998. Wall Street currently pegs the company's 1998 earnings at $1.24 a share, up from 89 cents in 1997.' So, the investor was already aware in February 1998 that the fantastic results of 1997 were maybe not so good after all because of the customers' debt and inflated inventories. He could have cut down his losses if he bought his shares at the offering, but why should he do so if the CFO of the company said that he believes that Mastoss will meet analysts' estimates for 1998? If they will do it or not, it has yet to be seen, but the minority shareholder is at least bewildered. Here he has a fantastic share with 32 consecutive quarters of spectacular increase, which he bought at $35 and thought it would triple to above $100 within a year or two, and he was going to sell it at a loss of a third of the price he bought it, just because the inventories have increased?

In the same month, on February 2, 1998, Mastoss announced that it had purchased Xovan for $35M in cash plus warrants. We had a year break, 1997, with genuine profitability that enabled the company to raise more than $100M at a valuation of almost a billion dollars, and then a year and a half after the acquisition of Furolias, Mastoss purchases Xovan. On March 3, Gommtow announced that it started coverage on shares of Mastoss with a strong buy rating. One of the top analysts says that at $22.875 stocks are trading at very attractive valuation. Gommtow sets 12-month $43 per share price target. Excellent news for the stockholders as Gommtow is a well known firm and if they give a strong buy recommendation at $22, they should buy more shares of Mastoss. Maybe, by buying more shares they will attenuate the loss that they have already incurred when they bought a few months ago Mastoss' shares at $35 in the offering. We will see later on the absurdity of the strong buy recommendations, which will be reiterated also when the shares' price will drop to $5. How can a serious analyst give a strong buy recommendation when the price is 20 or 30 dollars and still give the same recommendation when it is $5? After all why not, the one that foots the bill is the poor investor who buys all the rumors and recommendations available, and not the analyst. The analyst has no responsibility, his reputation is not tarnished in such cases, the management of Mastoss has no responsibility, they have declined all the responsibility in the prospectus advised by very competent lawyers. The buck stops ultimately at the poor investor's foot, as he loses his last buck…

A month later the shares' price is once again $28. And then, on April 27, the company issues a press release about the first quarter of 1998, with revenues up more than 25 percent from last quarter and operating income up 27 percent from last quarter, before Charges. In connection with the acquisition of Xovan, a one-time charge for in-process technology of $31M was recorded and, along with this acquisition, the company adopted a restructuring plan and incurred a non-recurring charge of $23M therewith. The loss including those charges was $44M for Q1 1998. Quite a fantastic loss, although it is extraordinary, but Mastoss has already shown how the extraordinary becomes ordinary, and this time they surpassed themselves. They paid for the company only $35M in cash plus warrants to purchase Mastoss' shares at $35 while the current price is much lower. The charges are this time about $54M, almost $20M higher than the purchase price… If the charge for in-process technology of $31M may sound reasonable for a company worth $35M, how is Mastoss going to expense $23M to restructure the company, in addition to the $31M? This amount of $23M is similar to the whole profitability for 1997!

Will the market buy such huge non-recurring charges? Will Ascorage, one of the big five auditors, buy it? It is not difficult to guess that they will. And if they do, how could we ever know what are the genuine operational profits for the operations of Mastoss prior to the acquisition, is this huge amount of $23M on top of the purchase price a shelter to hide operational losses that would be treated as restructuring charges? As long as the market does not penalize a company for losses, whatever are their origin, these methods will continue to prevail, and it will be legitimate, approved by the auditors, the SEC, the business community and the investors. A loss should be treated as a loss, extraordinary or not, for restructuring as well as from current operations. Is there a difference between a man who loses his sight as a result of a long sickness or as a result of an accident? In both cases he loses his sight and suffers the same consequences. At least the doctors do not treat the two cases differently, but they are not as sophisticated as the auditors...

This brings us to the recommendation to annul the term of extraordinary loss, in view of its negative impact on the understanding of the financial statements. All the losses incurred, charges for in-process technology, non-recurring charges for restructuring plans, and operational losses will be treated as ordinary losses above the bottom line. This does not mean that for taxation purposes the origin of losses may be treated differently, but in the financial statements disclosed to the shareholders all losses will be called ordinary losses. The shareholders can of course buy shares of losing companies at very high valuation, as we see all the time about high-tech growth company, but at least they will know what are the true losses of the company that will not be hid in legal terms approved by the auditors, but confusing the shareholders. Maximum transparency and simplicity, that is what this book advocates, as it is the best recourse for the baffled shareholders.

A week later, the share' price remained at $27. The market reacted favorably to the huge write-off. The illusion of the origin of the loss worked. In a market where you sell dreams and buy illusions, anything goes... In June 1998, the shares' price had returned to the price of June 1997. A full circle was made from $24 in 6.97 to $34 in 7.97, $25 in 8.97, $39.2 (the highest ever) and $36 (the offering price) in 9.97, the price dropping to $24 in 10.97. Curiously enough, the price dropped by one third only a month after the high price of the offering. Why is it so seldom the other way round? What miraculous forces bring up the shares' price just before the offering and let them drop sharply immediately after? And who pays for it – the underwriters who got their fees as a percentage of the high price's revenues, the analysts who gave a strong buy recommendation, the management who sold millions of dollars of their shares at the offering at the high price, or the poor investor who once again proved to be a sucker?... In November the price went further down to $22 and then up to $29. This fluctuation between $22 to $29 continued in most of the months until the price returned to $24 in June 1998. This brings us to another recommendation of a warning flag. Companies with shares' price that increase prior to offerings, when the controlling shareholders sell their shares to the public, and immediately after fall sharply, are suspected of unethical conduct and shareholders will abstain from buying their shares.

On May 30, 1998, we learn from the Internet that Gisco, another firm that was not familiar to the correspondents on the Internet gave on March 2, a sell recommendation, based on sales growth – low and down, higher receivables, falling gross margin, higher inventories. The price is still at $23. It is time to sell, to incur losses of one third and get out of this stock, but who ever heard of this firm, and besides they issued a neutral recommendation on March 20th. Yet, on May 15, 1998, recommendations of four of the top brokers of Strong Buy of Mastoss' shares appear in the Yahoo Finance Research, and no recommendation of moderate buy, hold, moderate sell, or strong sell. There was no mention of Gisco's recommendation.

On May 13, 1998, CF Rebel writes in the message board: 'I joked to myself that Nartokow's appearance on CNBC, wearing those proprietary erbium-doped glasses, was a ploy to ward off any potential acquisitor of Mastoss. We have a great management, period!... Mastoss is strategically positioned beautifully. Management has a strategic, long-term outlook – that's where the most bang for the valuation buck comes from. A takeover of Mastoss would be tactical only to this shareholder. Big deal if I got $43 a share for this great company tomorrow (my estimate as to what we'd be offered). The buyer would be getting off cheap. It would be a steal. Some would be tickled pink, now, to get this due to their past frustration. Well, where would you then put your money? You'll be extremely hard-pressed to find anything that would match the performance of this stock in the next 18 months. On a risk/reward basis, I contend, you'll find none. This company's stars are in syzygy.' This letter is so fantastic in view of what would happen subsequently, that one could ask is this gentleman is for real or is he a phony. He wants a takeover only at $60, at $43 it would be a steal, what will he do with the money, as this is the best investment in the stock market...

On May 22, 1998, Mr. Pink writes: 'Mr. Pink has shorted Mastoss in size for a number of reasons that shall be disclosed at a later time. For now longs should know that the sales staff at Mastoss are fleeing like rats from a sinking ship. So should you. Mr. Pink, offering a friendly warning.' On May 25, 1998, Idiot Detector comments on Mr. Pink's warning. He asks if there is evidence of his statement about Mastoss and if not that will be 'knowing misrepresentation'. 'We may need to build a case to take to the enforcement arm of the SEC that shows evidence of misrepresentation (based on posting of misrepresentation) and market manipulation (based on tracking of trades that show a pattern of price manipulation). And we need to get the officers and board members to pay attention and provide information to counter any claims that will be made.' We should pay attention to Mr. Pink, as he will be the hero of the following months and one of the first to warn the investors of the wrongdoing of Mastoss, advising them to sell their shares.

The investor will never know what are the motives of Mr. Pink. On the face value he states that he is short, so he is interested that the shares' price will drop as much as possible. Yet, if he knows some insider information and spread it on the Internet why shouldn't the minority shareholders benefit from it? Why should it be only the benefit of the controlling shareholders? The investor does not know at all at this stage if Pink is right or not, but he knows already that Mastoss has exaggerated in the write-offs of Xovan, there is a problem with the receivables and the inventories, and there is at least one unfavorable recommendation of a broker. Mr. Pink's warning comes on top of it. Mr. Pink is cursed by many correspondents. He is the whistle-blower, the one who troubles their peace of mind, their illusions. He is cursed like the prophets were for telling the truth. Nobody cares if his warnings are much more effective than the supposed protection of the SEC, the analysts and the auditors. He is motivated by his interests, whatever there are - profits or revenge, but this time his interests concur with the minority shareholders' interests. And when did the interests of the SEC or the auditors or the brokers concur with them? The auditors are probably interested in keeping their clients happy, the brokers are interested in gaining from the fees on the offerings paid by their clients, the interests of the SEC are not very clear, maybe they are submerged by work, maybe they are a bureaucratic organization that has forgotten its mission, but can the minority shareholders really count on them to safeguard their interests?

Seth Leyton calls Pink on the 22nd 'a liar and a S...'. Grommit writes on the same day: 'I am disgusted with the recent postings. With P around this thread has sunk to new lows.' Bruce L is more lenient to Mr. Pink and is not so sure if he is right or not as 'something is sending it (Mastoss) down'. mph says: 'I tend to ignore people like Pink. Too cute. If he has some hard facts, however, would be willing to listen.' Sector Investor is very ironic to Pink. He writes on the 22nd, the same date as Pink and as all the previous persons, 'Pink, I hope you shorted big! Please don't cover till after earnings – please, please.' Dee Jay writes: 'the only thing pink is your forked tongue'.

And then, three days later on the 25th Pink writes in the SI opinions: 'Mastoss is running out of money and burns a tremendous amount of cash, so a share repurchase would only accelerate the inevitable bankruptcy of the company.' And the same day he writes: 'Does anyone know if Nartokow is pals with ... Both are from Israel and both share a predilection for playing Chef Boyardee with their company's books'. He discloses, and yet we do not know if it is true or not, that the company is running out of cash, probably because of huge losses, and that the CEO is befriended with another Israeli who is ill-reputed for playing with the company's books (he means - tell me who your friends are and I'll tell you who you are). These are far more serious allegations than that the salesmen are leaving the company.

The Israeli connotation is very important, as the Israeli high-tech companies have an excellent reputation in Wall Street, as far as innovation, entrepreneurial spirit and technology are concerned. On the other hand, some of them have made a bad name for the Israeli executives and owners as far as playing with the numbers, selling their shares just before disclosing a sharp decrease in profitability, using insider information, etc. Many Israeli companies traded in the US have been sued, Soktow, Erinsar, Nalodo, and others, as explained in the previous cases, as well as many US companies founded by Israelis and operating in the US and in Israel, like Memnit and others. This ill reputation extends therefore also to US companies, like Mastoss, founded by Israelis (their origin is disclosed on the Internet but it also appears in the official reports of the company), and operating in the US and Israel as well. The Israeli connection rang a ring in the ears of many shareholders, who have not read it before in the reports of the company or on the Internet. Of course, this Israeli connection may sound anti-Semitic, as some of the letters will prove bluntly, but this book is not a sociological book and it deals only in economic implications of the disclosure.

This time, the reaction on Pink's allegations are much fiercer. Peacelover writes on the same day: 'This is a bold statement to make. Can you enlighten me on how you came about by this story? If your statements are not true, you better watch out, 'cause this baby is like a loaded spring. It can whack you real bad.' Pink fights back. Just to illustrate the speed of reaction on the Internet correspondence. The two letters of Pink were sent at 6.54PM and 6.59PM, and after receiving many answers he replies at 10.29PM: 'At their burn rate 50mn won't last long. Remember, Nartokow and the boys use fraudulent acquisition accounting to shift operating losses from income statement down to cash flow statement by overly aggressively writing off 'technology in development' when making acquisitions. They don't make any money. Read the income statements in conjunction with the cash flow statement. Maybe that's why the sales force is fleeing like high school kids from a cafeteria.'

This time Pink is even more specific. He probably knows the company very well, as it is almost impossible to figure out what he discloses from public information. He probably works or worked for the company, or has insider information from people working or connected to the company, Pink is also very erudite in accounting and very familiar with the stock market. He is an American, that we can learn from his style and idioms, No Israeli has such knowledge of American dialect, unless he lived for a long period of time in the States. He knows that his allegations cannot be proven legally and the financial statements of Mastoss are fully backed up by the respectable

Ascorage. But he knows what is going on, legal or not, it stinks, and ethically it is not permissible and works to the detriment of the minority shareholders. So he tells them: 'Run for your life as long as it is not too late. Look, you've been warned!'

Pink perseveres: 'Nartokow and the boys are making Falafel (an Israeli food) of the books. They exaggerate earnings by writing off nearly 100% of acquisitions (this is an understatement, as in Xovan's case they've written off 150% of acquisitions). That's a fact, ... and if you can't see that then you deserve what's coming to you. You might wonder also why it is recommended as a SELL Short candidate in two separate highly respected short publications.' We learn for the first time that Mastoss has also recommendations of SELL short candidate in two publications. That we did not know from the official publications on the Internet. The fierce answers force Pink to disclose more and more what he knows in order to prove he is right. WebDrone answers at 12.12AM (it is already May 26): 'As you point out with your incisive irony, Mastoss remains one of the best, most promising companies I can find. Now, if you really want to scare people, you are going to have to do a lot more research and analysis, and come up with something we have not already discussed and settled to our satisfaction. C'mon, this board needs a sharp, intelligent short with facts and articulated arguments – get to work! I'm the littlest minnow, and you are just making me smile. Try to make me worry, you goofball.'

At 3:31 a.m. (!) on the same day, Dee Jay writes: 'I believe Mr. Pink is a purveyor of slander of the worst sort, of innuendo with no basis, and of outright fraudulent remarks for some unknown (to us) purpose. At best he is a contrarian whose views should be seen as the reverse of the truth (got that, Pink? You know what I just called you?)' This rich dialogue hides a worry of the investors who are risking tens if not hundreds of thousands of dollars. One does not know who is in this arena the picador, the matador or the bull. One could only guess that as usual the owners and executives, the auditors and the brokers will not be hurt, and they let the poor investors play in the arena. The purpose of this book and generally speaking of ethics in business should be to make the companies, the affiliated/controlling/majority shareholders, the auditors, the analysts and the others enter into the arena and risk at least what the other investors risk.

Mr. Pink answers Dee Jay on the same day at 7:15 a.m. (those guys probably never sleep): 'Mr. Pink has been called worse. He remembers the insults when He told people to short …. (a list of companies in which his predictions were proved right). Your insults are like nectar. Remember in Plato's Republic how in the Simile of the Cave, the truth seeker is hated when he tries to liberate the cave dwellers from their ignorance and show them the light?' It is worth mentioning Pink's style as opposed to the other participants. Some of them use despicable language while Pink never loses control and shows that he is a literate man, who has read Plato, and sees all this campaign as fun while the others sweat and get angry. Meanwhile, the shares' price of Mastoss is still about $23, and the institutions ownership is about 40 percent of the shares. Tanners Creek is still optimistic and writes on May 26: 'When all the little investors (us) start crying, the big investors (them) come along and dry up our eyes and take the trouble (stock) off our hands because they are nice. Back up the truck and they will help you unload their trucks. We are at the bottom when I hear all this commotion. Moving on up. Got the guts to stay anyone? A good roller coaster (Mastoss) has to go down a little to move up. I love this ride and I'm hanging on !!!'

Other optimistics, as GenTechWriter, say: 'It's time to Plan to Persistently buy more Mastoss and Patiently wait for the market to realize what we on this board already know about our great company!!!'. One could not know if all those people are the same one, if they are long, if it is Mastoss itself writing the messages, and so on. The Internet is completely open to information, misinformation, slander, rumors, as all the writers are anonymous. Meanwhile, the Strong Buy broker recommendations have increased to 6 (!), which shows that it is very easy to make recommendations that will be proved in a short while totally unfounded when you do not incur any risk. Has any investor ceased to work with a broker after he made a wrong recommendation? Their annual EPS estimates for Mastoss was on May 21 - $1.25 for 1998 and $1.72 for 1999. One should understand the system that enables making such estimates that will be proved wrong. No sanctions are involved for the brokers and analysts, no sanctions are involved also for the company, as it has already made the offering at a very high price of $35 and even if the price falls to $5, it affects only the investor who wants to sell. The company, controlling shareholders and executives have raised more than $100M at $35. Those who foot the bill are only the minority (holding 80 percent of the shares) shareholders, that many of them bought the shares at $35 or more and want to sell them.

It is recommended to institute on the Internet, and later on issue a report by the activist associations, a pillory of the partners who have wronged the shareholders, without practically incurring any risk. Thus, the brokers, analysts and investment bankers who have recommended for companies like Mastoss, a Strong Buy and weeks later the shares collapsed will be published on the Internet, with their average rate of success in their recommendations. It is human to err, but it is not human not to suffer from it. So, if it will be disclosed that the less accurate company in its recommendations is Gommtow, the shareholders will no longer work with it and will disregard their recommendations. If among the investment bankers, Osttowar will be the worst company in its underwriting performance, and the shares of most of the companies they have issued have dropped substantially afterwards, the shareholders will no longer work with it and will abstain from buying shares from companies issued by Osttowar. If among the big five auditors, it will be disclosed that Ascorage had the worst performance in auditing companies that their shareholders suffered from irregularities in their accounting methods, the shareholders will no longer invest in companies audited by it.

In the same manner, the pillory will have a detailed list of companies, executives and directors who have acted unethically and/or who were convicted for defrauding their shareholders. If such pillory will be issued by a distinguished and objective body like ADAM, the shareholders will be warned in an adequate form and will be able to make rational decisions on investing into the company. It is obvious that no broker, banker, or auditor will ever have the highest note, but what matters is the ranking compared to the others, and the reminder of faults that were committed. Because of this pillory, firms like Gommtow will think twice before giving a Strong Buy recommendation, and if out of 100 recommendations 50 shares have dropped their prices immediately after, shareholders will treat them accordingly. If a car manufacturer has many safety problems, everybody knows about that, but firms that recommend buying shares of unethical companies are never penalized, although they wrong the minority shareholders who lose their money. Firms like Bonnty will think twice before underwriting companies with dubious accounting standards that can tarnish their reputation and the willingness of shareholders to subscribe to their offerings. The same cautiousness will apply to firms like Ascorage, which will be afraid that their competitors from the big five will win more ethical clients and that they will be left with the unethical clients, with more odds to be sued by the shareholders who will want to add to their lawsuits the 'deep pockets' of Ascorage.

The US market is a huge market and shareholders cannot keep track of all the companies and individuals that were accused of unethical conduct or convicted on felony. The activist associations and the pillory will have a very long memory and a fantastic data base and they will never forget, reminding the shareholders of the companies of the crimes and misdemeanors of the companies, executives and directors who have wronged them in the past. This pillory is not meant to become a vindictive vehicle, but just a yardstick on ethical behavior that will rank the companies, individuals and service providers according to their ethical standards. Today there are some buds of these ratings as has been reviewed in the book, there are ratings of the degree of corruption of countries, of the ethical conduct of large corporations and so on. But these are not prepared in a way that will give easy access and relevant information to the individual shareholder when he needs it, if he wants to buy shares in a public offering or in the secondary market. The Internet and the activist associations will give him free access to the information whenever they need it. Companies like Mastoss, and directors and executives of the company will hesitate before taking such huge reserves following their acquisitions and will be more cautious in their accounting practices. It may hurt the prices of the shares in the short run but in the long run they will benefit more from this rating, as well as their shareholders. In the meantime, shareholders will have to rely on information supplied by Pink and the like.

Pink continues to write on several boards. The prior ones were at SI, this one is at Yahoo, on May 26th at 11.37 p.m.: 'Israeli management team proves once again to be criminal. How else do you explain 31 consecutive quarters of earnings yet, negative shareholder equity. These guys come to our country because they know their native countrymen are too savvy to be duped by these crooks. They don't make money. Losses are hidden in cash flow statement under write-offs of technology. Sales force is fleeing and OEM customers have dumped Mastoss for more reputable vendors. Watch this bad by tank a doodle doo. Nartokow is just like ..., the crim.' This time, Pink goes too far, he has hurt the national pride of the Israelis and the Israeli community which is very strong and influential in the US is offended and he starts to get offended answers, like this one from Iowewian on May 27 at 1:26 a.m.: 'Hope to buy any share you may be selling tomorrow, Mr. Pink. History of Mastoss and Israel is against you!!! Take heed, Mr. Pink to the last verses of Amos. Amos 9: 14-15, 'And I will restore the captivity of My people Israel. And they will rebuild the ruined cities and live in them, they will also plant vineyard and drink their wine. And make gardens and eat their fruit. I will also plant them on their land, and they will Not again be rooted out from their land which I have given them, Says the Lord Your God.' Quite an awkward statement from an Israeli who has left his homeland to make money in the US to quote Amos who promised that the Jews will not be rooted again from their land. But what do you expect at 1:26 in the morning?

But apparently Pink is also optimistic, as he advises on the same day 'Jump in when the stock hits 12-13. That's when Mr. Pink will be buying – to cover His short.' Pink probably has also his 'air de grandeur', as he calls himself Mr. Pink and refers to himself as 'He' with a capital letter. Probably, 'He' should be compensated for his self-esteem after all the curses he receives... He is reminded that 'many pigs are pink'. He is threatened that he is 'probably being investigated by the SEC right now for stock manipulation and fraud'. It is very easy to analyze in retrospective what has happened to Mastoss and pity the shareholders who went to Canossa to Pink after the shares fell to $5. But we have to try to figure out what was the state of mind of Pink in those days, with all the curses and threats, what strong character he needed to answer and to fight back, his drive was probably beyond the profit he was about to make from being short on Mastoss, it was maybe revenge but maybe also integrity and the ethical motivation of preventing wrongdoings made by the company to the shareholders. Pink sounds sincere when he writes to Lucrative on June 2: 'The stock market Gods are just. Apologies to Homer.'

On the same day we learn of the French connection of Nartokow, the CEO of Mastoss. TexasCanuck writes: 'I enjoyed the review in IBD, noted that Nartokow got his MBA in a French institute which should make him familiar with Europe.' And on June 1 and 2 Pink continues his crusade stating: 'More sales people are leaving. So should the shareholders', 'The share price has dropped about 20% since Mr. Pink first issued His warning about this stock. Mr. Pink, He does not lie.' E pur si muove, He does not lie... And then, a coup de theatre, Mortishko, one of the founders, buys 10,000 shares and Nartokow buys 5,000 shares. This shows confidence in the company at such a difficult stage, until Edward Leinbach mentions on June 2 that in September '97 they sold 200,000 and 100,000 shares respectively. 'So this recent purchase is nice but may be more cosmetic than anything. I'm pulling for Mastoss, that's for sure, but this seems like almost a non event.' The problem with the Internet is that you are completely transparent, and the curtain is raised from all your attempts. It was a nice try, but the sting was taken out immediately afterwards. The skeptical shareholders believed that a personal loss of a few hundred thousand dollars for a public relations fireworks display is rather too much, it could have been better to hire at this price a PR manager or agency that would have assisted Mastoss much more in those troubled times.

On June 8, Seth Leyton makes a very pertinent remark, commenting on the conduct of the underwriter of Mastoss, who he thinks was supposed to support the share that they have issued: 'It gets so frustrating watching ... do nothing but sell. If your mechanic never fixed your car, you'd go somewhere else. If your doctor always hurt you, you'd find another, but if your investment banker is nothing but a seller.... You do nothing? A poor analogy, I know, but come on. Why is it impossible for ... to find buyers?' But the shares' price is still $22.75, the clouds are gathering but the price does not move. In an average volume of 550,000 shares per day, there are still shareholders that have confidence in the company and buy the shares, after all, $12M trade in a day with no fluctuation of the price is difficult to achieve if everybody sells. So, who buys? Probably, the unsophisticated investors who do not follow the thread, or the overoptimistic who do not want to be confused with facts, or the innocent believers in the financial statements and the auditors who cannot be wrong.

And on June 15, after quoting Pink from another thread (Pink was active in several threads) who stated: 'This stock will probably drift along until the holders realize that Godot is not going to show up. Mr. Pink, a martyr.', randall c. cummings reacts: 'Look for more malicious lies, rumors, innuendoes, distortions, and other shenanigans from this low-life short & his colleagues and lackeys. Only this time, I think he will get his head handed to him on a platter.' The reactions are getting more and more fierce, Godot and

Salome are recruited for help, especially since the share on the same day opened below $20. On June 16 at 5:54 a.m., peacelover, the same one who threatened Pink very bluntly, raises a Machiavellian suggestion that Pink is being backed by Mastoss who wants to go private: 'Once the Pinks have driven this stock to say 9+, then Mastoss will take this co. private and screw us shareholders. Has anybody thought of this? Maybe we are dealing with the most unscrupulous management that is intentionally driving this stock down. Otherwise, I can't figure their non-chalance.' The poor investor is now completely baffled, if peacelover no longer believes in the company, what will happen to the share?

But two hours later, Pink answers: 'It is a silly theory. They are desperate to hype the stock. Unfortunately their rumored buyer bought … instead. Kind of like being the last guy at the party and all the girls have gone home with somebody else – unless, of course, if you are short like Mr. Pink.' WebDrone, another fan of Mastoss, answers peacelover, a few hours later: 'If I thought management was anything but honest, I would just sell at any price, and walk away, You do what you want.' This is interesting, for the first time honesty is mentioned. Until now the investors in the thread sounded like credulous speculators, no one talked about honesty, integrity or ethics. And then, all of a sudden, a categorical statement that if the management is not honest, this investor is going to sell at any price. He is willing to incur a heavy loss for ethical reasons, that is new and completely unexpected. Pink or the others have just to discredit Mastoss' management, and even if the stock is attractive, there will be some shareholders who will sell their shares. How will the believers in Profit Superstar react to investors who are not only motivated by profits but have ethical considerations as well?

And peacelover perseveres, on the same day at 7:20 p.m., he writes: 'The leaders of Mastoss are arrogant buffoons and if they don't get their acts together they will have their butts kicked by the likes of the pinks. Ain't that a shame, unless of course they have some ulterior motive for their actions or lack thereof. Peace and good luck.' The investors started to lose their politeness only after the provocation of Pink. We noticed that in Erinsar's case they remained polite while they lost most of their investment, but then in Erinsar's case everybody kept his coolness, especially Astossg. In Mastoss' case, because of Pink's allegations made in blunt terms, the investors lose their sangfroid and adopt vulgar expressions like 'they will have their butts kicked'. Where has the respect that they had for the three musketeers of Mastoss disappeared to, when the fiercest remark made on their personality was on the spectacles of Nartokow. And it is only the beginning, as the shares' price will drop more and more, the remarks will be fiercer, with anti-Semitic connotations, with personal allegations, that are among the worst ever written on the Internet.

Comments like 'You idiot, go die', 'I won't take s… from the likes of you', 'I'm pretty sure the big dogs at Mastoss read this board', become common. Randall c. cummings continues to attack Pink and writes on June 22: 'I encourage others to also voice similar complaints to SI. This gutter slime must be stopped. He is no better than the scum who cries fire in a crowded theater. There are limits to free speech. There are also such things as morals, ethics, honor, etc.' Once again, reference to morals, ethics, honor, etc. but tied up to a call to limit the freedom of speech. And for what? Even if Pink spreads totally unfounded rumors, who can assess it. Could SI be the arbitrator between investors in the hundreds of threads? The essence of the Internet is full freedom of speech, especially in economic matters. Ethics is mentioned with coercion, honor with shutting the mouth of a potential benefactor of the investors. But then, Pink was right when he quoted Plato in this respect.

And Pink, like a flea, continues to bite. On the same day, he writes: 'Mastoss = Accounting Fraud.' 'Nartokow was full of his usual hype and b.s. Investors are going to want Nartokow to show them the proverbial money. But this laggard in the industry is going down.' 'Mr. Pink is not alone. …. have both called attention to Mastoss' deteriorating position, and aggressive accounting. Escape before it is too late.' This time, he gives the names of the firms that gave a sell recommendation, and after all he is no more alone. He was the first one, but others are rallying to his fight. And he warns once again the investors 'escape before it is too late'. On the other hand, one could imagine how Nartokow could be furious from Pink's allegations, calling him laggard and full of b.s. Yet, Pink is not afraid of fighting Nartokow, a multimillionaire, who could sue him if he manages to break Pink's anonymity. Randall c. cummings continues inexorably to attack Pink. In June 29 he writes: 'Sorry, Pinky or Porky or whatever, you have to gain respect in order to be treated respectfully. You obviously have forfeited any possibility of respect by your past behavior (lies, deceit, manipulation, etc etc).' And Pink becomes more specific. He foresees the fall in the third or fourth quarter of 1998.

More and more discussions on the integrity of Nartokow and the other executives are being held, are they or are they not crooks, that is the question that preoccupies the investors. To that they add, are we going to sue or not. But the general impression is: keep quiet, don't sue if you do not want to affect the shares' price, the management of Mastoss is honest and speak to the investors who call them and are answered personally(!) with integrity. Others quote their brothers who are working at Mastoss and give them tips about the earnings per share. They are warned not to give insider information, if their information was genuine anyhow. But others, as bc 1111 on July 19 are happy to receive insider information: 'You must be a real jerk to have reported someone on this board for sharing information. I hope you are not serious. If Peter's info is correct, then I'm glad to hear it, illegal or not. This is America, home of the free.'

And in the middle of this rumors, we learn by Webdrone on July 19, 1998, that Mastoss has raised through Paroll '$100M (!) in cash through a bond offering giving 5% for three years and conversion to shares at $27.5.' All Pink's prophecies about cash burning rate became irrelevant as Mastoss acquired cash that could help them survive for a long period of time even with substantial losses. In retrospective, should Mastoss' management have not sold those bonds in a brilliant move at the eve of the catastrophe it could have collapsed and Pink's prophecies would have been proved totally correct. But with $100M in cash on top of the existing reserves the company could face serious setbacks without fatal repercussions. Which is to the credit of the management and owners that they knew all over the years to rescue the company from all the dangers encountered in this high risk business, whether by acquisitions, doubtful write-offs, offerings of shares or in extremis sale of bonds.

Like always this was done at the expense of the poor shareholders who bought the shares at their peak or bought the bonds just before the collapse of the shares. Yet, nobody sued them and no investigation of the SEC was conducted. Nartokow who studied in a French University could have told his partners the old French proverb: 'Les chiens aboient et la caravane passe'. The Internet had some nuisance value, made some noises, but did not change any substantials. Or maybe one, a most important one, the sophisticated shareholder who read carefully all the financial statements, press releases, prospectuses, and now the Internet threads with warnings of Pink and others, have for the first time enough data to make the right decision. He has of course to discern genuine information from misinformation, but there are no free lunches, and before that he did not have even the rough data. So, there is a substantial progress, not enough, but the Internet and whistle-blowers' warnings like Pink's are indeed a step in the right direction.

Meanwhile, the Pinks start to multiply, or is it Pink who writes to himself? On June 29, PaperChase (a new one) writes to Pink, at 3:58 a.m. (this thread can be called the red-eyed thread): 'Pinky. The $53 million in accounting charges last quarter are going to hide the sins of Mastoss for sometime. Are you willing to wait 8 to 12 months for the big fall? It may take even longer. The restructuring charges are total B.S. and will hide the real losses for a while. I'm sure they wrote off everything including their wife's underwear in those charges.' Valuefinder writes on July 2: 'EXIT as fast as you can from this scam. DO NOT BUY into this GARBAGE. They are doing some creative stuff away. This is a time bomb ready to explode. I know, I have been there before. TIC... TIC... TIC...' Lucki678 gives the heading of 'Ex-Israeli army' to his message of July 7: 'Shareholders? Management runs co. like a commando operation in hostile territory. Expect surprises but no revelations. Anyone got an infra-red scope?'

VladTepesz, a new comic hero of the thread, writes on July 8: 'Somebody pay attention to me! I deserve your attention! I'm wearing a tutu and dancing for you, why won't anyone pay attention to me? Boo! Scary! Scary! Sell your stock! Pay attention to me!' And he continues a day later: 'Everyone on these boards tends to be rah rah cheerleaders, and here we are, hoping for the stock to drop – can't blame us for trying. We know that our posts are useless, but if you respond to what we post with such enthusiasm, well, you have only yourselves to blame... With today's little move, I am going to accept that Mastoss is not likely to reach my target price of 10 and move on. Worst of luck to you all.' And bennythebug, his counterpart, answers him on July 9: 'Vlad: you seem to enjoy sending messages under other names. It's nice to play but I think it's time you found another to play your games on. Who knows, you may even find another little boy who wants to play with you.'

These comic interludes emphasize the tragicomic character of this case. It just depends where is your position in the play. It is a comedy for the controlling shareholders who are laughing all the way to the bank with tens of millions of dollars raised at the highest valuation. It is fun for the auditors, the brokers and analysts who receive their high fees regardless of the valuation. It may also be amusing for the SEC officials who have their pensions secured, regardless of the outcome to the 'speculators'. It may even seem ironic for the writer of this book who looks at the events of this case with a benevolent eye and a lot of compassion for the losers. But it is undoubtedly tragic for the minority shareholders, holding 80 percent of the shares, who are about to lose shortly, as we shall see, hundreds of millions of dollars of valuation when the shares' price will drop from the twenties to five dollars. People are going to lose their savings. We shall see the tragic connotations of the collapse and how it affects normal people who are not at all speculators but who wanted to share with the success of a remarkable high-tech company.

The absolute order of magnitude of the losses in this case, hundreds of millions of dollars, is what differentiates it from the other cases. The wrongdoing in the other cases was much more flagrant. Here, we have at worst a creative accounting that has been backed up by the auditors and not blamed by the SEC. We have operational losses that maybe were hidden under huge reserves of the acquisitions and were disclosed at last in 1998, but we do not have a selling of the company to affiliated or new parties in extremely unfavorable conditions to the minority shareholders, without letting them have the opportunity to share in the future of the company in which they have invested. In Mastoss, the controlling shareholders still retained 20 percent of the shares of the company, and they did not sell the company, as they tried all by themselves to turnaround the situation. It is beyond the scope of this case to analyze the reasons for the shares' price' fluctuations after their collapse in 1998. It is irrelevant for the shareholders who had to sell their shares in 1998 and lost up to 85 percent of their investment. Yet, if we were to rate the players in this tragicomedy, the Mastoss players are the least unethical yet. We do not know what will be the future of the company, and if they have learned something from the very negative response of the stock market to the wrongdoing they have allegedly made.

Or, if encouraged by the fact that nothing has happened to them or to their company, no lawsuits, no SEC investigation, and no personal losses, as they did not sell their shares immediately before or after the collapse; they will try the salta mortale, as the Erinsar's owners did, risk 'le tout pour le tout' and sell their company to the detriment of the minority shareholders. The experience of several Israeli entrepreneurs, whose companies have collapsed and whose shareholders lost most of their investment, shows that after a few years, they ride again, make successful IPOs of new companies and new suckers invest into their new companies. In some cases the class actions were settled for a few millions dollars, which is only a small percentage of the shareholders' losses, paid by their companies, but they personally were only marginally affected as they managed to jump off the sinking boat before the collapse of the shares' price.

Vlad continues to make his predictions in his comic way and it is still unclear if Vlad does really exist or is he the comic alter ego of Pink. He writes on July 9: 'There's just no telling how low some people will sink. Forget the imitations. Just listen to the real thing. Me. Vlad. And what I'm telling you now, you can take to the bank. Tomorrow, Mastoss is down. You heard it here first – none of this stupid hiding behind phony names. Shorts rule. Longs drool. Get a life people.
Virtually
Living the
American
Dream'

And then, on the same day, in a sharp contrast to VladTepesz's humorous posts comes the first anti-Semitic message written by SteveHide: 'Buy buy buy you know the game, Mastoss, the Zionistic conspiracy continues into … Mastoss will rule the world. ….. don't have a chance. We will avenge against the Eurotrash Nazi world and dominate ... DEATH to he Aryan... We will rule. Remember Treblinka..' The investors knew for some time of the Israeli connection of Mastoss. When everybody was winning money nobody made any allegations on the management's origin. But when the first clouds begin to gather, we are reminded of the renowned 'protocols' and of the Nazis and anti-Semitic allegations that the Jews want to rule the world and the Aryans have to fight back. From then on, the genie is out of the bottle and the posts will get worse and worse. This is also another facet of the Internet allowing such people as SteveHide to write freely. Has anybody the right to stop him?

Does it have any impact on the investment community? It is too early to tell, as the Internet is in its infancy stage.

In this tense atmosphere, on July 10 at 1:02 a.m. we read a message sent from djane to Sector Investor stating 'Mr. Pink revealed' about an article of the Wall Street Journal called 'Individuals Caught in Crossfire Of Duo's Battle on Cosham': 'Florida stockbroker Dostier and his wife recently fled their home for five days to avoid a barrage of obscenity-laced phone calls – including death threats - that he says was unleashed by an adversary who goes by the moniker "Mr. Pink." It's just one chapter in an odd and disturbing tale of small-stock investing on the Internet era. "Mr. Pink" and Mr. Dostier, using his screen name "Skipard," have been slugging in out for months on Internet message boards dedicated to the stock obscure Cosham Mr. Dostier, 56 years old, is a big stockholder and head-over-heels fan of the money-losing New York company, which is trying to market a device to diagnose jaundice in newborns by analyzing their skin color. "Mr. Pink," a Cosham critic, describes himself as a hedge-fund manager who has sold shares short in a bet that the price will fall. Caught in the crossfire: lots of individual investors wondering which – if any – Internet guru to believe. "Mr. Pink" and his fans have been the big winners lately, as Cosham shares have collapsed. The stock closed at $5.50 Thursday in Nasdaq SmallCap Market trading, after closing as high as $17 on April 30.

Battling in the apparent anonymity of cyberspace, "Skippard" has slammed "Mr. Pink" as "pond scum" and a "slimeball". "Mr. Pink" labeled Mr. Dostier "a loser" who "led your followers to ruin". Then "Mr. Pink" escalated the feuding by posting Mr. Dostier's full name, employer and address on the Internet. "Anyone upset about losing millions on Cosham should call Dostier at (his home phone number). Or send him a package with a gift to (his full home address), advised a June 22 posting on Yahoo! Finance. "I am a big boy when it comes to winning and losing money in the stock market," says Mr. Dostier, who works from his home as a broker for .... But faced with personal threats, he says, "I'm afraid". While many of his longtime followers haven't denounced him, he says, lots of other unhappy Cosham investors "now have someone to blame". The Cosham message boards are evidence that some of the virtual communities spawned by the Internet are downright ugly places, driven by hostility and awash in adolescent name-calling. "Skipard" and "Mr. Pink" are like the captains of two teams engaged in a frenzied brawl.

The on-line war of words over Cosham also highlights the risks for investors at the dangerous juncture of small-stock investing and the Internet: This new medium has made it far easier for adventurous stock pickers to compare notes. But it also helps wrongdoers spread misinformation to pump up the price of a thinly traded stock they hold – or drive down one they have sold short. Both sides on the Internet debate on Cosham have leveled allegations of market manipulation against the other, and a lawyer from Cosham says U.S. Securities and Exchange Commission staffers "said they would follow up" on the company's complaints about possible trading irregularities by short-sellers. (Short-sellers sell borrowed shares of a company they bet is overpriced, hoping to make a profit by buying cheaper shares later on). An SEC spokesman said the agency doesn't comment on pending investigations... The company's attorney declined to say whether the firm is considering legal action against "Mr. Pink"..

In recent weeks, Mr. Dostier says he has been swamped with phone calls and e-mails from panicked followers who say their retirement-account balances, their financial futures and even their marriages are in jeopardy because of huge bets on this one speculative stock. Many had invested heavily in Cosham shares using margin loans and have watched their brokerage firms liquidate shares to satisfy "margin calls" as the stock price fell. Mr. Dostier says he feels responsible for his followers' losses because they looked to him as a "rabbi" – a role he clearly relished in happier days. But he also says, "They shouldn't have bought so much stock. They shouldn't have margined it." Cautionary messages have rarely showed up in Cosham postings by "Skipard", however. He has repeatedly talked about his own huge bet on Cosham – now over 400,000 shares... "What first caught my eye about Skip was his passion," says ..., an Illinois homemaker who posts on the Internet as "Janybird". Ms. ..., 36, who looks to Mr. Dostier as a mentor, is now suffering with 80% to 85% of her family's investment dollars in Cosham. "I still think I made the right decisions investing in this company. I really do," she says. But the stock-price collapse "has been a killer"...

Internet posters have offered several names as the possible identity of "Mr. Pink," including Leon, 36, managing member of hedge-fund firm Treon in New York. Mr. Leon, in an e-mail response to several messages, said, "Treon does not comment on the existence of its short positions and does not comment on negative stories about companies." He didn't respond to the question of whether he is "Mr. Pink" in his e-mail, and said he didn't have time to respond to questions in an earlier, brief telephone conversation from his office. Mr. Leon's firm has certainly put its money where Mr. Pink's" mouth is: Of a sampling of 18 stocks offered as "picks" by "Mr. Pink," 14 showed up as Treon holdings on March 31, according to Treon regulatory filings collected by Technimetrics Inc. Treon's top four holdings ... are stocks that have been strongly recommended by "Mr. Pink". Treon's record suggests the followers of "Mr. Pink" have been right to hang on his words: The firm's primary fund returned 44.3% in 1996 and 52.5% in 1997, according to Managed Account Reports, New York.

"Mr. Pink," in the phone interview, asked a reporter not to "out" him as he outed Mr. Dostier. At the same time, he expressed no regrets about disclosing

Mr. Dostier's identity and accepted no responsibility for the harassing messages Mr. Dostier says he has received. "I didn't make a single one of those calls," he says. "If people did that, they were wrong." Mr. Dostier says he "would pay a lot of money" to learn the real identity of his Internet adversary – but says he wouldn't send "Mr. Pink" anything beyond a Christmas card. Above all, he says, "I just want my vengeance in being proven correct" about Cosham."

Apparently, the Mastoss thread is becoming very interesting. Now it has also suspense! But this brilliant article raises many questions. Is it possible that Mr. Pink is just gathering some information on companies, true, false or half true, spreads rumors on the Internet, causes systematically the prices of shares to fall and then gain huge profits from its short positions. And even if it is true, if his information is correct, as it is apparently the case in Mastoss' cases, what wrong is it if he gains huge profits? The fact is that Mastoss' shares collapsed a few months after Pink's allegations as he forecasted rightfully that they were not earning money and Mastoss incurred indeed operational losses for the first time in many years in spite of the huge reserves in Xovan's acquisition made a few months earlier. But even if he is correct in his predictions does he have the right to get 'personal' by disclosing the phone number and address of the alleged wrongdoers? What if a lunatic sends a bomb to the address that Pink has disclosed, could he still say that it is not his fault? On the other hand the personalization of his opponents to the angry shareholders can deter potential wrongdoers from doing their schemes. If the fight becomes personal, maybe the immaculate affiliate shareholders will think twice. And what if they are lynched? When does a legitimate fight of minority shareholders stop being correct and becomes dirty? What if there are no other alternatives?

The writer of this book is convinced that in no case the fight is allowed to become dirty. It is never to become personal and affect the personal lives of the wrongdoers and their families. It has to stick to the rules of politically correct, without any allegations to the religion, gender or color of the shareholders' opponents. An ethical fight cannot adopt unethical means in order to win its case, and of course no unlawful means are to be adopted. This conviction stems from ethical reasons as well as from practical ones. When the fight becomes dirty, the odds that the wrongdoers will retort using harsher means is very strong. If a shareholder discloses the personal phone number of his wrongdoer, immediately afterwards he will be himself the victim of disturbing phone calls, his family will be in jeopardy, and his life may also be in danger. The majority shareholders and the companies are always stronger than the minority and will always win in dirty fights. The minority shareholders have therefore always the necessity to keep the rules of fair play even if their opponents retort to foul play. They should not be like Gandhi, they can be tough and give a good fight, but ethics and justice have always to be on their side in order to win the fight in the long run.

And then, following djane's disclosure on Mr. Pink's name and allegedly wrongdoings, Steve Sohn answers her on the same day with another scoop, that the Court has decided to raise the curtain on people who have made negative comments about a firm on the Internet. "Philip Pierce's Net Secrecy Court decision gives beleaguered company access to names and addresses of people who have made negative comments about the firm on Internet chat group – by The Financial Post – Philip Services Corp., its stock decimated by a barrage of writedowns and troubling accounting practices, has quietly won a court order forcing about a dozen Internet providers to cough up names and addresses of people who posted negative comments about the firm on an Internet chat group. The move has potentially chilling implications for privacy and the Internet. It means Canadians who exchange information and opinions in chat groups have lost the traditional cloak of anonymity and can be held liable for what they say. The order, granted by Ontario Court Justice ... in Hamilton, was made ex parte – without Internet providers including America Online Inc., AOL's CompuServe division, iStar Internet Inc, and Westling Datalink Corp., being notified or present to make arguments.

It instructs the providers to hand over to Philip names, addresses, e-mail addresses, telephone numbers, computer serial numbers, and other information for a specific list of messages posted on Yahoo in April, May and June. It doesn't stop there. The providers were also told to preserve "all other messages sent by such persons through the Internet providers." And they were ordered to supply Philip with the real identity of the users who posted messages under pseudonyms – common practice in chat groups. Philip was granted leave to examine the information, although that decision was later reserved pending another hearing. The court also ruled that the files be sealed and expressly forbade the company, its employees and agents to "publish, speak about or distribute this order or any documents provided with the order." Many of the messages, which can still be read, appear to make allegations of criminal activity against Philip executives and express fears of what might happen to anyone who exposes too much about the firm's activities.

But Philip spokeswoman ... said it was company employees who felt threatened by what they were reading. That's why Philip decided to act. She said some of the worst messages have now been pulled by Yahoo at Philip's request. "The tone of the board became increasingly malicious and downright defamatory." ... said. "It libeled employees of the company, issued threats of stalking, a whole range of ethnic slurs, and got to the point where employees were very concerned. So the company decided it was going to take action." This article could have worried much the whistle-blowers, as the Mastoss' share price on that day was still $23, and the events did not seem to prove that the allegations of Pink, VladTepesz and the others against Mastoss were correct. They were indirectly threatened that they could be sued as in Canada, their anonymity was no long taken for granted, they could be prosecuted in justice, their name and phone numbers could be made public, and what Pink did to Dostier could be done to him. Following a potential court order or the article disclosure on the identity of Pink, it would be possible to disclose Pink's phone number as he did to Dostier, people could phone him at his private home, threaten his family, and make his private life miserable. He could be no more immune behind his pseudonym.

Very optimistic messages followed, bennythebug was confident that he will enjoy the ride up based on the $1.75 earnings estimate for 1999 and that Mastoss should be trading at $35-40 by the end of 1998. Mr. Pink becomes more specific and predicts on July 20 that in the third quarter of 1998 the fall will occur. Math1000 retorts on July 22 that 'after looking at the charts on Mastoss for the past several months, I really believe that without any negatives surprises next week, the stock will go up to about $43. JUST MY OPINION from doing the math and looking at the charts!!' The mathematicians of the thread are the most ridiculous of all, like astrologers or chiromancers they take their charts and know for sure that the share will be traded at $43. In fact it fell to about $5, but what arrogance to find scientific justifications to hunches that might be correct or not. And many shareholders who are ignorant on those matters are impressed by their self-assurance and buy the stocks. Before the fall there was a league of most of the thread to convey the message that all is calm on the western front of California and that Mastoss will continue its ride to the top of Disney World. Only Pink, VladTepesz, and occasionally someone else presented a divergent opinion and they were immediately crucified as the enemies of the people.

Roktar attacks personally Mortishko one of the founders and Mastoss' Chairman of the Board. On July 24th he writes: 'Mortishko is probably a great scientist and entrepreneur. As Chairman of the Board however, he should be voted down for failing to represent the interests of the shareholders.' In the last few weeks, the shareholders are getting nervous and attack more and more the management personally. And on the same day the underwriter of Mastoss' 1997 offering, Bonnty reiterates its BUY rating to Mastoss. Asked why they raised the $100M convertible debentures a few weeks ago, Mastoss' management says that they want to have the immediate financial flexibility to make future acquisitions. But Texas Hillman is still worried on the same day of the timing of the new debt. And he states one basic axiom that should be written on the walls of every Board room: 'MANAGEMENT NEEDS TO SERVE THE SHAREHOLDERS. WE ARE THE OWNERS OF THE COMPANY, AND THEY ARE OUR EMPLOYEES.' It is so true, and yet so out of context as the immediate future will show.

And on July 30 Gommtow reiterates its STRONG BUY recommendation based on the second quarter's results that beat their estimates. Mastoss reported EPS of $0.31 exceeding their estimate of $0.29. The revenues increased to a record of $66M, the 34th consecutive quarter of sequential revenue growth of the company. Net Income was $8.7M. And they finish their July 28 comments by stating: 'We believe that Mastoss shares continue to be very attractive based on valuation. Mastoss shares currently trade at modest P/E multiples of 16x our 1998 EPS estimate and 11.5x our 1999 EPS estimate, a significant discount to the long-term earnings growth rate of 40% and at a discount to its ... peers. We are continuing to maintain our 12-month price target of $43.00 and our Strong Buy investment rating. On June 26th, the company completed $100 million 5% convertible subordinated note issue, which raised net proceeds for the company of $96.4 million... Management indicated that the proceeds would be used to pursue an aggressive foray into ..., as well as opportunistic acquisitions. Overall headcount at quarter end was 723."

But these results do not impress Pink. On August 1 he writes to Saul Feinberg Jr.: 'Saul, you seem to know a lot about the company, but know nothing about reading financial statements. This company has serious quality of earnings problems. It makes no money and burns cash by the ton. It will go lower no matter how much technical sounding mumbo jumbo you recite.' The term 'technical mumbo jumbo' is so accurate to this case and to many others as analysts rely too much on technical analysis and forget about the basics of common sense, accounting, psychology, integrity, ethics and so on, which contribute much more to the company's valuation than the bottom line numbers. One has to read behind and above the bottom line, one has to understand the personality of the managers and the owners, what motivates them, what is their record, in order to understand the company and predict its long-term growth. And yet, the shares' price on that day is above $20, as on the one hand we have the anonymous Pink and the buffoon VladTepesz and on the other hand we have two large Wall Street firms as Bonnty and Gommtow. Pink, the intellectual philosopher or the greedy speculator summarizes the situation on Sunday August 2 by quoting Dante's Inferno: 'Abandon hope all ye Mastoss shareholders'. So who is right?

And on the same day, waltzing in the ballroom of the Titanic a moment before hitting the iceberg, dobr assures that the salespeople were NOT 'fleeing like rats from a sinking ship', he knows that this ship will never sink and sends a message to the perplexed shareholders stating: 'Well, brilliant engineering, great fundamentals, huge market and 7 years of solid 40% growth history are enough for me... All of the negatives are clearly fictional, and have created one of the greatest opportunities I've ever seen. Anyone still afraid of Pink's "one-liners" should have this all clarified now. A year from now we will all remember this and have a good laugh (while those who sold because of Pink's rants will have a good cry). Good luck to all.' And later on, Jack Colton praises Pink in one of the most well-balanced message of the thread: 'Actually, I find Mr. Pink's views something to keep in mind. He has actually challenged everyone on the thread to examine Mastoss in intricate detail. If we cannot find something wrong with Mastoss ourselves, to substantiate Mr. Pink's views, then we are better off – and our investments are more sound. But if we were to find merit in one of Mr. Pink's claims, we would be well advised to take caution.'

And this is the crux of the matter. In spite of his allegations, Pink did not succeed in lowering the price of Mastoss' shares that remained throughout his messages in the low twenties. The only reason of the collapse of the shares is Mastoss' announcement that they will not meet their targets. Only substantial facts influenced the prices, not rumors, as the US market, unlike the Israeli market, is very sophisticated, the average trade volume of Mastoss' shares was about half a million shares a day, or about ten million dollars, and no Pink or ValdTepesz could influence it. Only hard facts, strong evidence, has brought the collapse of the shares and therefore the shareholders have to praise Pink who was the only serious prophet who predicted the collapse, the reason and the exact timing. They were warned soon enough to sell their shares at a loss, but not at a huge loss, as some of them did a few days later. This is the advantage of the Internet, you get the raw material, and you have to scrutinize it and decide what is genuine and what is not. Which is a lot better than the situation that prevailed a few years ago.

Yet, a very strange event happened in August, moments before the collapse, serious investors, like Mr. Green, started to post the thread, saying that they were large firms with many customers, and assuring that they were continuing to buy the shares in the twenties as they saw Mastoss as an excellent investment opportunity. Green recommends to the shareholders on August 1: 'expect new and positive coverage on Mastoss within the next two weeks' and assured on August 3rd that his firm's 'investment plans in Mastoss run into 8 figures'. It is very suspicious that such posts were spread especially in this timing, stating that serious firms planned to invest tens of millions dollars in Mastoss. It could be that large firms, learning on time from insider information of the forthcoming announcement of Mastoss, took the opportunity to divest their preferred clients from Mastoss' shares on time. We remember the miracle of Apollo in the Furolias case who had the insight, that had nothing to do with being a member of the Durtem Group as Furolias, to sell all his Furolias shares a few days before they collapsed. Such miracles might have occurred also in Mastoss' cases, as the trade volume was quite high and the messages were too optimistic. This of course did not bother the SEC, the auditors or all the other institutions that were supposed to safeguard the interests of the minority shareholders who did not have any insider information.

Albeit all the allegations and attacks, Pink does not lose his temper. He reiterates on July 28: 'Mr. Pink never claimed to be an authority on ...; he does know accounting and fraud. Mastoss' accounting, while probably not criminal, is aggressive and overstates the company's performance.' And on the same day he writes: 'Mastoss is a scheme bordering on fr**d. Where is the cash? Earnings are made up and occur only due to accounting gimmickry. Mastoss is a bad accident waiting to happen. Did you miss Mr. Pink's calls on ....? And with the gentlest blow from Mr. Pink's mouth, Mastoss shall topple like a house of cards. Mr. Pink, He speaks the truth.' What a panache! But the fortunetellers start to multiply. On the same day Mr. Busdriver predicts: 'Hop on the bus folks, we're going down... NEXT STOP....' Davidraziel writes on the 29th: 'There is something fishy going on here. I would buy more here, but I am having grave doubts about the straightforwardness of management.' And brim41 attacks personally Rostronsky's, one of the founders, ultraorthodox Jewish background on July 28th: 'BEWARE DON'T TRUST THE CO-FOUNDER BAAL TSHUVA. HE LIES WITH HIS EYES'. HellMan5 writes on August 1: 'The real reason Mastoss won't go up. I've finally figured it: management is overly greedy.... They want to grow faster, make more money, take over more companies, become big players overnight. Too much greed is a bad thing (even on Wall Street). IMO management is starting to behave like megalomaniacs. Here's my message to management: MELLOW OUT.'

We remember what happened to some of those Wall Street yuppies who were too greedy and how ethical management preaches exactly the opposite. When the shares' price skyrocketed and increased by 1000 percent no investor of Mastoss complained about the alleged greediness and acquisitions policy of the management. When the shares' price ceased to climb and even decreased by 40 percent, the investors started to speak about ethics and now about the greediness of management. Which is good, as morality's birth is primarily on difficult periods. When greediness pays, very few want to be ethical, but when the situation deteriorates everybody joins the Salvation Army. As it is the strong belief of the writer of this book that the wrongdoing to the minority shareholders will ultimately lead to the collapse of the wrongdoers' companies, it is only natural that the shareholders would change their attitude before it is too late and in this way prevent unnecessary chaos.

And while the shareholders are getting more and more nervous and some of them state that they have sold their shares at a loss in view of the rumors, Pink writes on August 10: 'Mr. Pink loves the smell of Mastoss getting pummeled in the morning.' And the shares started to slide slowly in August and closed on the 10th at $19.25. And on August 19 Pink tirelessly writes: 'Please explain how Mastoss has had 32 quarters of growing earnings but has never generated cash and has not increased book value. The proof is in the shareholder equity pudding. Mr. Pink gives Mastoss the highly coveted Triple T rating (triple turd) for aggressive accounting that could be overturned by an SEC that has found religion on this issue.' But neither the SEC nor Ascorage, the auditors, did respond. On the 21st when the shares trade at $17.75 Pink writes: 'Mr. Pink's friends in the Israeli community tell Him that these guys are scum bags. The Wall Street Journal today had an article about these manipulative sham write-off transaction that are abuse to inflate earnings. Get a clue, get a life, read the writing on the wall. Mastoss is going down big time as soon as the SEC forces them to stop making up their numbers. Wrong stock to own in a bear market.'

Richard Birecki, who apparently knows personally the management of Mastoss, answers Pink on August 21: 'Is the "Israeli community" so small that your Israeli friends know Mortishko, Rostronsky, and Nartokow personally? Mortishko does one thing, WORK, that's all. He is always in some foreign country closing some deal. I'd assure the same for Nartokow and Rostronsky. Outside of the ..., I really don't think that many people know these three cause they have little time for their families, much less socializing. Even if these guys ARE scum-bags, don't you realize that they have too much at stake to be dishonest and send their stock down... Please explain why people with 90% of their net worth in this stock would risk everything and cook up numbers? Please tell me the INCENTIVE behind it. I think we can all agree that people who take 20K to a ½ billion $ company are not stupid, please explain... I won't go into detail, but I know for 100% personal fact that Mastoss leaders want to see the stock go up, for their families as well as themselves.' Assuming that Birecki is genuinely a shareholder and not a cover-up of Mastoss' management, he is an example of the self-deceit and the innocence of the unsophisticated shareholders. He overlooks the fact that the three managers have already cashed out tens of millions of dollars at the highest valuation of the company with a minimum dilution.

If it is correct that they have started the company with $20K, it is a remarkable return on investment. Of course they would like the growth to continue forever and that is not their interest to bring down the shares' price. So the questions of Birecki are completely irrelevant. He should have asked: 'How come the shares' price was at their peak precisely when they sold their shares and since then the price has only decreased? And if they lose from the low prices it is only a paper loss, as the only thing that counts is what is the price that you pay for the shares and what do you get when you sell them. In this respect, the management has not lost anything and had a fantastic return on investment, while the minority shareholders are about to lose up to 90 percent of their investment if they sell their shares after the collapse.' This is exactly how Istovius earned also millions of dollars from his shares and warrants in Soktow and Erinsar, Gorekius earned millions of dollars from his shares in Erinsar and Memnit and even earned money from the sale of his Furolias shares, while the minority shareholders in those companies have lost most of their investment in the events described in the Furolias, Erinsar and Soktow cases. According to Istovius' law only one thing counts, at what prices are the executives and owners getting their shares and at what prices do they sell them. And the sophisticated shareholder should use this information, which is made public, in order to buy and sell his shares.

In case that the reader of this case might get the wrong impression that the Mastoss case is an isolated case that happened a few years ago and does not reflect the tendency in Wall Street, the thread quotes on August 24, 1998 the following article: 'SEC Considers Limits on Acquisitions Write-Offs', by The Wall Street Journal. "The Securities and Exchange Commission said it may impose new limits on the way companies book their acquisition write-offs due to the poor quality of some recent corporate earnings reports. In particular, the SEC is worried that a rising tide of companies are abusing acquisition charges for purchased research and development, goodwill and restructuring costs. Accounting critics say some acquiring companies are reporting dubious write-offs for these costs to artificially 'manage' subsequent earnings. 'How auditors report these charges in financial statements doesn't change the value of these companies,' said ..., accounting and tax analyst at Bonnty. 'But the disclosure and transparency of these charges should certainly be improved so investors and analysts can properly assess the value of these companies.'

Banof, director of the SEC's division of corporation finance, said the SEC first has to discuss these problems with corporate and accounting executives. Once that is done, the SEC will then consider 'tightening the accounting rules or auditing rules' for these charges or 'we have our own rules that we might change', moves that could take place next year, he said. The SEC enforces accounting and auditing rules. It has the legal authority to enact such rules on its own, but rarely does so. 'We are considering our regulatory options at this point,' Banof said. Notably, a growing number of companies are writing off huge chunks of their acquisition costs as purchased R&D. The higher the value for these charges, the more acquirers can avoid hits to future earnings from goodwill. When companies purchase other firms, they must write off any resulting goodwill, the premium paid over the fair-market value of an acquired company's assets, for as long as 40 years, slicing into earnings every year along the way. A recent New York University study shows only three companies wrote off part of their acquisitions as R&D during the 1980s. But 389 have done so in the 1990s with a record 156 in 1996 alone...

In addition, the SEC is worried about the rise in reported restructuring charges, Banof said. Such charges are usually taken for things like layoffs or plant closings, and can temporarily depress profits, but make earnings glow in subsequent years. They are 'often characterized as one-time unusual events, but if they are one-time unusual items, then why are they becoming more usual?' Banof asked. The SEC's signal that it may tighten the rules covering these write-offs comes in the wake of the disclosure last week that the agency's Office of the Chief Accountant is meeting with accounting and corporate executives to discuss the recent wave of corporate accounting problems. Over the past year or so, more than a dozen companies, including ... have had accounting problems blow up in their faces. The SEC's heightened concern stems in part from its perspective that auditors aren't doing enough to stop companies from bending the accounting rules to manage their earnings. Auditors increasingly are asking the SEC to bless certain dubious accounting practices that their corporate clients demand. 'We won't go away, we're here and we're vigilant,' Banof said."

The minority shareholders of the third millennium can be secure, as the SEC has promised that 'we don't go away, we're here and we're vigilant'. The future buyers of shares from offerings underwritten by Bonnty do not have to worry, as 'the disclosure and transparency of the acquisition charges should certainly be improved so investors and analysts can properly assess the value of these companies'. And after the SEC will 'discuss these problems with corporate and accounting executives, tightening the accounting and auditing rules' there is no doubt that companies like Mastoss and auditors like Ascorage will show penitence and write a book on business ethics. This article shows the enormous difference between the theory and the practice in real business. The problem is not if the companies succeed through creative accounting to postpone tax payments. The IRS is very mighty and can take care of itself. The tragedy is that the companies are tempted to hide behind acquisitions and restructuring charges operational losses, thus postponing the disclosure of the losses until they manage to raise new equity at enormous valuation, selling owners and management's shares at those valuations, and when the true losses have to be disclosed a year later the only ones who suffer from the collapse of the shares' price are the minority shareholders. And all the bodies that have to protect those shareholders speak highly of reforms but do very little, if at all.

A few days before C day (collapse day) Pink speaks of himself as almighty, after all on the Internet you can be anyone you like. He writes on August 23, when the shares are still traded at about $17.5: 'Oh, you do not seem to be aware that not only does Mr. Pink control a vast financial empire, he sits at the helm of a sprawling network of associates, affiliates, operatives in the field, confidantes, agents, advisors, consultants, etc. that feed him information virtually 24 hours a day.' Even if Pink operates only by himself in an attic in the Bronx, he will prove a few days later that he was smarter, with more foresight and information than all the underwriters, brokers and analysts of Wall Street working in Manhattan in billions of dollars organizations, covering Mastoss for years and giving it a Strong Buy recommendations hours before C day. One could almost believe Pink when he describes himself on the same day and a day after: 'Mr. Pink is a good Samaritan and is here because He loves the common man and wants him to avoid financial tragedy. A good Samaritan enjoys unmasking criminals and exposing fraud for the benefit of the public at large. Not to say Mastoss management are necessarily crooks, per se, it is just what Mr. Pink likes to do to serve society.' Si non e vero e ben trovato…

BOGEY MAN phones Mastoss on August 24 and is assured that 'things were looking good for Mastoss in the near future.' One could ask how near is near, as a few days later will be C day. It is like seeing the light at the end of the tunnel, but the light is of the train that is going to collide with you. And when the tension arises a 'comic' interlude of VladTepesz on August 24, responding to his opponent bennythebug: 'I vil destroy za zionist dogs of za … vorld. Ve vil prosper and have the Arian corporations of … destroy zose Israeli pigs. Yes, mein lieber, it iz your medication time again.' And he continues a day later: 'Sorry if you were personally affected by WWII, but I don't have any sympathy for the Israeli gov't considering their actions in their own country. I don't want to get into a political discussion or a discussion about the ethics of satire.' Here again the ethics are being quoted in the name of free satire, or 'comic' anti-Semitism. On August 25 Black September continues: 'Ask the Israelites how many a clear sign have We given them, and whoever changes the favor of Allah after it has come to him, then surely Allah is severe in requiting (evil).' And iowegian replies on the same day: 'What are you advocating? Driving them into the sea. Maybe another holocaust? Seems to me that Allah is giving you a clear signal that you just won't listen to. Love your neighbor as yourself. Do unto others as you would have to do unto you.'

And if the anti-Semitic remarks did rally the Israelis/Jews to back Mastoss in its difficult moments, why not rally also the Europeans. Iecut quotes on August 26 a certain Roger de Belgique who is really amazed by the way the Americans treat Mastoss' share which is 'highly valued' in Europe. He writes in French, which is very rare in the US as the Americans understand very rarely foreign languages. Of all people it is VladTepesz who translates almost perfectly the message six minutes later: 'I am really astonished to see how the American public forsakes this value. It is warmly recommended in Europe. It is mentioned in the larger financial publications of Europe.' Iecut continues a few moments later quoting the same Roger in a message called 'It's now or never': 'La valeur est vraiment sur-vendue et c'est le moment de faire le plein des achats. En Europe, cette action est conseillee par les grandes banques et on ne comprend pas pourquoi le public americain la boude.' The perplexed American investor who buys his perfumes and haute couture from Europe should therefore pay attention to Roger and buy Mastoss' shares, because the largest banks of Europe do so. Older an Wiser writes a few moments later: 'It is a Jewish Company. Nobody will merge with them....they openly refuse to have anything to do with a Jewish Company.' And HenryReardon replies: 'Mastoss the chosen company. This stock cannot suffer just because it is Jewish. Ever notice how light the volume is in the market on Rosh Hashanah and Yom Kippur?' And Black September terminates this chat saying: 'This is not religion, it is the prophecies... The children of Israel will suffer from their betrayal.'

Meanwhile, the suspicions of the serious shareholders are getting stronger and stronger. Sheldon Feinstein writes on August 26: 'I must say I have never seen anything act as irrational as this stock has, unless there is something going on that we do not know. I am coming to the reluctant conclusion that there is something going on and that the shoe will drop and we will see the stock, at a minimum, cut in half when this happens. I have been assured by people who supposedly know that there is nothing going on and all is well, but I am beginning to believe their information is false. I defy anyone to explain the action of this stock, over the last year or two, in light of the numbers, unless the numbers are being cooked. I hope I am wrong.' Sheldon, did you sell your stock on time or did you stick to it until the bitter end? On August 27, the stock price falls by 15 percent to $14.375. Snoopdaddy1963 writes: 'What a BUMMER!!! Fear is Near! Is Mastoss a steal at these prices'. VladTepesz writes: 'Sniff sniff sniff...' Bennythebug retorts that he is a messenger of Mephistopheles sent to make sure he makes as many people as miserable as he can. He advises him to return to Hades and tell them that he did a great job above and beyond expectations. Gobuffs98 is fatalistic and writes about his bad sentiment, summarizing 'BUMMED OUT IN Mastoss LIFE GOES ON'. And Mr. Pink summarizes on the day of the fall, a few hours before the announcement: 'Mr. Pink prefers the low hanging fruit – outdated technology and criminal management – a lethal combination.... Look for further drop to single digits.'

The first one to advise the Yahoo trade of the collapse was of all people, bennythebug, the opponent of VladTepesz who predicted weeks before C day the collapse while benny strongly supported Mastoss' management and prospects. On August 27 at 9:47 p.m. EDT he writes in a heartbreaking message: 'BAD NEWS guys. Mastoss just warned on the 3rd quarter. 10 to 15% lower than expected sales, which means about 55 to 60 million. Lower earnings than q2 and about 8 to 10% higher expenses than q2. New products will be delayed until the end of q4. I guess this means we're in trouble and shorty prevails. Someone had to know this was coming.' Someone knew, dear friend, and you did not want to hear him, you ridiculed him, you cursed him, you had eyes but you did not see, ears but you did not hear, nose but you did not smell that something fishy was going on. You have nobody to blame but yourself, but still you are not responsible, you were only innocent and fell into the trap that most normal shareholders would have fallen in. It is unbelievable that an ethical company can publish excellent results of the second quarter, convey messages to the analysts and shareholders days before the warning that everything was fine and prospects were excellent, and then all of a sudden you tell them a month before the end of the quarter that you will not meet by far the forecasted results. It may be legal, but it is not ethical. An ethical and responsible management that cares for their partners and shareholders would have warned them much earlier and not keep their illusion that growth will continue forever.

On Thursday August 27 1998 at 8:40 p.m. Eastern Time Reuters announced: 'Mastoss sees lower Q3 revs, income vs Q2. Mastoss said Thursday it expected the third quarter to have lower revenues and income than the second quarter due to weaker demand for its ... and delays in introducing the next generation of ... Mastoss, whose ... run more efficiently, said in a statement it expected revenues to be 10 to 15 percent lower than the $65.7 million reported in the second quarter. It also said it projected lower income than it had earlier anticipated but did not provide any figures. First Call's consensus estimate from analysts who follow the company forecasts an operating profit of $0.32 per diluted share for the quarter against $0.23 for the same period a year ago. For the second quarter, Mastoss reported a net profit of $0.31 per diluted share against $0.21 for the same period a year ago. The company said in the statement its gross-margin percentage would likely be down from the previous quarter's 44.1 percent... Mastoss said operating expenses are expected to be about 8 percent to 10 percent higher than the $17.4 million reported in the second quarter. "In addition, delays in the introduction of next generation ... products that were expected to be released before the end of the third quarter of 1998, contributed to a shortfall in Mastoss' revenue expectations," it said. "The company's plans now call for introduction of those products before the end of the fourth quarter of 1998," it said."

How can we construe this warning – is it a normal event in the history of a high-tech company that after 34 miraculous quarters' growth, we are forecasted that the sales will be 10 percent lower than last quarter and operating expenses will be 10 percent higher, or do we have to attribute to the company a scheme meant to fool its shareholders? As the methodology of this case is to stick only to information made public on the Internet, we shall try to answer this question by analyzing the response of the market to this scoop. Bennythebug has recovered from his initial shock and an hour after his pessimistic message he returns to his original optimistic tune. Definitely, there are some blessed people who will remain optimistic even under the harshest conditions. He should have taken the pseudonym of Candide. He writes at 11:01 p.m.: 'Time to pull up our guts! ...Lets hope that they can resume their exciting (I think that was the word Nartokow used) growth in the q4 period and continue on the right track going forward. We will probably need a few words of encouragement from management about the future... We are still a growing company in an exciting industry. Management just has to stay up with their game plan... It is easy to get down on Mastoss but almost all business is suffering around the world... I don't think this is the time to throw in the towel. Keep your chin up and let's try to encourage management instead of taking potshots. They brought us this far let's hope they can score big with our new products. Once again I want to reiterate this news hurts me as much or more than it hurts others. But I am hanging in, I hope you will all join me for the pot of gold that awaits at the end of the rainbow.'

He almost said that the shareholders were responsible for the collapse... In the Diaspora, before World War II, there were many Jews who were convinced that if there is so much anti-Semitism the blame was on the Jews. Maybe they were too successful, maybe they were too different, maybe they had really a despicable character and an ugly look as the pamphlets described their stereotype. Even, in the state of Israel, supposedly deprived from the inferiority complexes, the President said immediately after the Yom Kippur War and the fiasco at the beginning of the battles: 'we are all guilty'. This school of thought believes that the wrongdoers are not to blame, they find attenuating circumstances to their conduct, hope that they will mend their way, it is not time to throw in the towel, keep your chins up, we need them, 'I love my master', as the slaves in the Bible who did not want to be freed said and remained in slavery. Fortunately in the US, at the end of the millennium, those bennythebugs were the exception and most of the correspondents understood the events in their true meaning.

SJANDREWS writes a few minutes later: 'Mastoss - Management Really Very Corrupt. Well guys, here you have it. The end of the long dusty trails of BS accounting and BS talk of ... this, and ... that. The wheels have been coming off this dog for over three quarter now, and if you are still long then you are still wrong. Selling now is like shutting the barn door after the cows have left, but at least it will get you back to cash and out of this crap piece of paper. If you want any of your money back I suggest a lawsuit, against Bonnty and Paroll. They both just completed a private placement of this junk. Sorry that you are left holding the bag, but I can't say you didn't deserve it, given all the warnings. Asta la Mastoss.' This gentleman reminds the shareholders that the warning came immediately after the offering of the convertible debentures that helped Mastoss raise $100M, the highest amount ever raised by the company, he reminds them of the warnings, of the irrelevance of the talks of the management about technology and new products, and advise them to sell their shares even at the low prices.

And on August 28, at 2:21 in the morning (anti-Semitism never sleeps) shauls write: 'Damn Israelis. I warned you guys over and over don't trust Israelis. They are sneaks and dishonest, each and everyone. Particularly the Ba'al T'shuva with the deep-set dishonest eyes. Good Shabbos.' Once again, a direct attack on Rostronsky, the founder, and this time also a defamation of all the Israelis. What is really hard to understand is how this shauls who knew that the company was run by Israelis did ever invest into Mastoss. We know that there are funds in the US that invest only in Christian companies. It may not be politically correct, but it is understandable, so why doesn't shauls invest in them? Maybe because those funds do not give enough return on investment, maybe because most of the Israeli stocks quoted in NASDAQ gave a fantastic return on investment on the average. So, Israelis/Jews are good only as cash cows, but when they lose money, they become sneaks and have dishonest eyes? Those anti-Semites, speaking on behalf of honesty and trust, would have probably lynched the Israeli management, dressed as KKK members, ethically...

VladTepesz also does not sleep on the same night and he writes at 2:09 a.m.: 'I truly am sorry about the bad news... I kept saying the management isn't worth a crap. Well, enough said. Good luck and I really do hate to see anyone lose money/equity.' This book will not quote messages with very dirty language, but it is important to tell that a lot of them followed the collapse. From the parts of the message of davidraziel of that night that can be quoted: 'Now we know why they were so desperate for the cash – acquisitions – bullshit!!, there is going to be little cash to the bottom line from operations and Nartokow needs a new pair of bifocals. As for ..., I hear that they were the P.R. company that generated the Memorandum to the Titanic crew re: arrangement of deck chairs. Good night fellow victims.' The absurdity of the thread is that it brings to very strange coalitions and shauls uses identical curses (c...) of those of davidraziel, who said in a former message that he was Israeli, and tries to convince him that all Jews and Israelis are the same c...; all that at 2:45 a.m..

On August 24, three days before C day Zacks disclosed in Yahoo Finance that there were three Strong Buy recommendations, no moderate buy, no hold, no moderate sell, no strong sell. Mastoss was ranked second out of 81 in its industry. The estimate of EPS were $0.31, in consensus, holding as 90 days ago. The shares' price was then about $18. Four days later, after the announcement the price collapsed to about $6, a 70 percent decrease. The volume of trade on August 28 was more than ten million shares, compared to an average of 439K shares. If we remember that there were altogether 26 million shares, that the owners held about 5M and did not sell, and that many Americans were on holidays and heard of the news later on, we can conclude that in this day more than half of the shares available in the market were sold. Who bought those shares, we shall never know, was it the owners who took the opportunity of the collapse to buy indirectly their own shares as was alleged on the Internet, was it institutional investment, was it new shareholders that seized the opportunity and would make a 400 percent return on investment after less than one year when the shares reached the twenties in

1999? It appears that the poor shareholders who sold their Mastoss shares on 28.8.98 did not have any other alternative, as most of them were answering Margin Calls and were forced to sell. This is another tragedy that if you buy on margin and the shares collapse you have to sell incurring all the losses, without having the opportunity to stay with the company until the storm is over, if at all.

Bestphotographer recommends on C day that 'the management ought to be sued for over this'. How could they be sued if they received the best legal and accounting advice and were acting with no legal default. Cerahas pursues: 'Where is the class action lawsuit?? The way I look at it, I have already lost my shirt, now I just want to hurt Nartokow in any way possible.' There were no class actions on Furolias, there was one lawsuit against Loskron and the minority shareholders lost and were sentenced themselves to pay indemnities, there are many class actions on Erinsar and Soktow pending but with little chance of success. In Mastoss' case, class actions had a priori no chance as the company acted within the law, although unethically. Which reiterates once again the difference between the law and ethics, and how the law cannot safeguard in most of the cases the interests of the minority shareholders. Those shareholders can complain and cry to no avail, as Ibthinkin on August 28: 'Nartokow the idiot... He lied about the future after the last earnings claiming that everything was looking good... Mr. Pink had the last laugh after all on the Silicon Investor thread where all the fools believe everything Nartokow says like the gospel truth. At this point I'll hold because I don't have much more to lose. I do however think that a lawsuit will probably be the shareholder's only recourse. I wonder if Nartokow bought a summer home on the French Riviera like they say the Russian politicians/swindlers did?'

On the air stood the allegations, like those of robt123 on August 28: 'I'm concerned that the insiders knew they were going to suffer these setbacks when they sold millions in stock several months ago', or kuggle's 'My opinion is that the company had to or did some kind of insider trading, because how does a company come out with earnings and not foresee a problem in revenues and a slowdown in producing the newer tech. I strongly believe that I was mislead by their last conference call.', or securitiesP1: ' It is a fraud run by a bunch of dirty bastards who liked to cook the books a la ... This explains why they had such a great earnings report recently. What they did was push product out the door and move up sales by starving future quarters. That gives you a blowout quarter and the attempt is to pump up the stock so insiders can bailout before the truth comes out. With creative accounting you could easily cover it up.' The management was blamed of the collapse with allegations such as: 'You have lost credibility, you are a disgrace, the mushroom theory of managing this business (keep everyone in the dark and feed them sh-t), this type of management capable of being caught with their pants down in the accounting department'.

And there are always the smart guys like wonderabout who boasts on the same day: 'The warning signs were there! When this stock did not go up on the last earnings announcement, I felt there was something amiss. I sold at $21.75 and rebought today at $6.125. Surely, this is a bottom fisher's paradise.' The last word of this memorable day is to Mr. Pink who jubilates in his victory: 'You fools should have listened to the wisdom of Mr. Pink. He does not lie, unlike Nartokow and the analysts and foolish bulls on this thread that have led you to ruin. Hopefully you have learned a lesson. Where there is aggressive accounting/fraud as in the case of Mastoss, it indicates serious underlying problems. Anyone who wants to pay homage to Mr. Pink may do so at His thread. The rest of you may kiss his well-toned bottom.' And even Saul Feinberg Jr. pays his tribute a while later by writing: 'Let's give credit where credit is due. Part of becoming a successful investor is accepting defeat. Mr. Pink was right.' While, after reading the complaints on his rude language, Pink apologizes: 'Sorry about the fools comment. And sorry about the financial losses. Mr. Pink has suffered much rudeness and even threats from certain participants on this thread and He should not have grouped you all together.' His Majesty is after all only human...

Barbara J. Payne praises Pink for being generous in sharing his opinions with the shareholders and philosophies: 'The stock market is for big boys and girls, not for those who wish to blame their misfortunes on those who do accurate analysis. It was up to you to verify or discredit his assertions. Apparently, you thought that unnecessary, and now you are paying the price for your lack of diligence.' As mentioned before in this book, the people who come with an original truth are first of all mocked and ridiculed, than despised and cursed, and at last praised and taken for granted.

After a fantastic ride, Mastoss returned to where they started it three years ago in 1995. Within a year prices climbed by 1,000 percent from about $4 in spring 1995 to about $40 in spring 1996, after a drop in price it reached once again about $40 in summer 1997, and a year later in summer 1998 it returned back to $5. In the meantime the insiders have earned tens of millions of dollars, some minority shareholders had big wins and some lost 80 percent of their investment. To illustrate the impact on one individual shareholder we quote the letter of signist, an old timer of the thread, to his daughter Lorraine on August 28: '...I am very sorry I talked you into investing your hard earned money in this company as the Management of Mastoss lied to my face when I visited the company last year when I asked them if they planned to support shareholders. In fact, they don't give a damn about their individual shareholders and have not been creating a respected company purely by their hard, honest work. No, I can't sue them personally for this, however I wouldn't be surprised if a class action lawsuit would surface soon...

This is a sad day for me especially because I involved you in this company. It still has large amounts of cash (duped from new investors recently under the guise of the need to make more acquisitions...) and if you hold the stock for approximately 6 months to a year you probably will get your money back. We will have to talk about how you might still be able to buy the house you have been looking for as obviously the money you counted on or I made you believe you would have with your investment in Mastoss will not be available to you now. You are aware that Your Mom and I have significantly invested in this company and worst we probably will have our margined (borrowed money) called (demanded to be paid) immediately. While at this time, this situation will create a hardship on us, we will be ok... I think. This has definitely changed my outlook for this company. It may regain it's stock price but will never repair my disappointment with being involved and invested in this company.'

The letter personalizing the individual shareholder touched many correspondents to 'think good thoughts' for him, to have the highest regard for him, they were sorry to hear of his misfortune, and were glad that he shared the letter with them. After all, a friend in need is a friend indeed. Only Pink reacted, as usual, cynically, on August 29: 'Your story would break Mr. Pink's heart, if He had one.' The others suggested to signist to send the letter to Mastoss' management. If he did, they probably were very touched and compensated him fully for his losses, as Istovius did to Karisios in Furolias' case. After all, goodness is contagious and has also permeated from Erinsar to Mastoss by the acquisition of Furolias. Or, on the other hand, Mastoss' management could have thrown signist's letter away and said: 'Ah, another speculator crying with crocodile tears, after all, we have lost much more than him, instead of winning hundreds of millions of dollars from our initial investment of $20K, we made only tens of millions.' And indeed they were in a very bad shape with a market capitalization of $194M. More and more threats of legal actions against them were appearing on the Internet, as Bill Z Ridley PA who wrote on August 31: 'Anyone interested in pursuing a class action lawsuit against Mastoss please contact me directly .... I am looking for a lawyer.'

On August 29, after his 'victory', Pink bids farewell from the thread, to those who threaten him and those who congratulate him: 'You have learned that it is bad luck to threaten Mr. Pink with legal and regulatory action... This is Mr. Pink's last post here. His job is done. For now He bids you all adieu and says that He wishes that He received the same respect He showed all of you. He does appreciate the few words of thanks He has received mostly privately. He prays that you learned an important lesson in all of this. Good Bye, Mr. Pink'. Everything is related, and before separating from Pink we read a post of starpopper on the same day addressed to Pink: 'I see you went to Tel Aviv University. Does that mean you had a grudge to grind with Mastoss/Furolias?

They were an Israeli company before getting bought out ... could you be a disgruntled EX-employee?' Is it indeed possible that Pink was a disgruntled Furolias employee, that was bought by Mastoss two years ago. But how did starpopper learn that Pink had studied at Tel Aviv University, and even so, many employees of Mastoss were Israelis, so he could be a disgruntled Mastoss employee, or he could be connected with disgruntled employees from Mastoss or Furolias. We will never know, but it does not matter, as it emphasizes the main issue of transparency. As long as Mastoss is transparent and is not afraid of disgruntled employees they should not be worried of information spread on the Internet. Only when a company hides information, plans schemes, engages in creative accounting or uses insider information, only then the Internet can endanger a company. Transparent companies, like honest people, are not afraid of anything as everything is in the open, and the Internet is a threat only to opaque companies.

On C day, the 28 of August, Bonnty, the underwriter of Mastoss and one of the largest investment bankers of Wall Street, issues an announcement on Mastoss, analyzing the recent events. They lower their 1998 EPS estimate to $0.85 from $1.25 (the actual was $0.53), and their revenue estimate to $245M from $270M (the actual was $264M). The 1999 EPS estimate was updated to $0.90 instead of $1.75 and the revenue estimate to $315M. And they conclude: 'We are maintaining our long-term Buy rating on Mastoss shares.' Full stop, no apology to all the investors who counted on their research and bought the shares because they issued a buy recommendation. The investors lost in a few days 80 percent of their investment, and Bonnty's recommendation is still buy. It was Buy when the price of the shares was $40, buy when it was $20 and buy when it has collapsed to $6. It is amazing what high level of credulity of the investors is needed to take seriously such recommendations from a 'creme de la creme' firm. It makes one despair on when will they ever learn. On Monday August 31, Nartokow held a conference call, giving an impression that they could write off the next two quarters, and having no concrete feelings till '99 when new product will be developing and testing. As ntdy has cleverly put on the same day: 'Mgmt's conservative style is serving them well in this case. I think you probably have more luck suing investors for 'general panic and irrational behaviors' than suing Mastoss.' On September 1, Gommtow announced that it maintained its Strong Buy recommendation, yet lowering earning estimates...

The circus still performs, the fool's ship still sails, and the shareholders receive the same type of information. In a few days management changes its forecasts from one extreme to the other and nothing happens. They still have the support of Bonnty's and Gommtow's analysts who recommend 'buy' at whatever price of the share and never 'sell'. They have the support of the auditors, the respectable Ascorage, the same auditors of Furolias, Nalodo and Erinsar, who approve of their accounting practices. This case tells the story of the individual shareholders, who have sold more than 10 million shares in one day. Half of the equity has changed hands in a day, people have lost hundreds of millions of dollars in a day, and nobody cares. Most of them were obliged to do so because they had borrowed money to buy the shares and were obliged to sell them, after they collapsed, at the request of the banks.

Business is as usual, a few complaints, a few cries, Lorraine will not be able to buy her house, some shareholders have learned a lesson, Pink jubilates, some anti-Semitic gentlemen have one more reason to hate the Jews, and nothing changes. The shareholders will continue to invest in shares of other companies, after all you win some you lose some, they will not cease to invest in companies that are suspected of unethical behavior, they will not invest in Ethical Funds, they will not leave forever the Wall Street market and invest in saving accounts, they are still lured by the American dream, they want to become millionaires at all cost, even at the cost of losing all their savings, but what are savings in comparison to dreams, to illusions, to make believes?

Unless repetitive cases of scams, full disclosure in the press, the television, the Internet, books and theses, will cause an ethical revolution of the minority shareholders, that will force full transparency and ethical conduct of companies, through the leadership of activist associations. These vehicles will prove to the stockholders that there is an alternative way, that if they unite, if they protest, if they cease to buy shares of unethical companies, they can make the change, not in a year but within a few years, before 2010, until a new regime will prevail, the democracy of ethical companies, owned by most of the citizens of the country, managed by ethical managers, and supervised by impeccable ethicists.

12
CLASS ACTIONS

"The problem with being in the rat race is, even if you win, you're still a rat."
(Lily Tomlin)

Since the case studies mention very often class actions, it is necessary to clarify this term. This is done in a very detailed way in the site of Schubert & Reed LLP on the Internet, which gives information on the legal terms and practices of the suits that are available to the individual shareholder. These legal vehicles, mainly the class actions, are very limited in their scope, rewards and efficiency. They are time consuming, and some people even alleged that they benefit mostly the lawyers who handle the cases. Still, until more efficient vehicles are devised, many shareholders resort to class actions.

Securities Fraud Class Actions

If you purchased shares of a publicly traded company trusting that the marketplace had properly valued the stock based on the complete mix of information available, and it later turns out that the price you paid for your stock was inflated because the company was dishonest about important adverse information, or hid such information, and the price of the company's stock goes down significantly, you may be a member of a defrauded class of shareholders, eligible to pursue a class action claim under federal securities laws to recover your losses. A class action claim for securities fraud may be brought by an individual plaintiff on behalf of all persons similarly situated, or by a group of representative plaintiffs.

These cases are brought with the assistance of experienced class action counsel, who will bear the costs of prosecuting the case and apply for a fee, to be approved by the court, only upon the successful conclusion of the case. Any settlement of a class action case which would bind the members of the class of injured persons who are not represented individually by the absent class members, must be approved by the court as fair, reasonable and adequate.

Shareholder Derivative Actions

Claims for breach of fiduciary duty typically arise when a company's directors and officers cause the company to violate the law, exposing the company to criminal or civil penalties, massive losses, and to damaging litigation, such as securities fraud class actions. The fiduciary duties owed to a corporation by its directors and officers include the duties of due care and loyalty, and require directors and officers to obey the law (and cause the corporation to obey the law). No director or officer may seize a corporate opportunity for personal gain without giving the corporation the chance to take full advantage of that opportunity. No director or officer may place his or her personal enrichment ahead of that of the corporation. Nor may a director or officer simply abandon his duties to the corporation.

When a company has been wronged by its directors and officers, and might well sue them and recover, a dilemma exists: the very people who have damaged the company and who should be sued are running the company – "the foxes in charge of the hen house." In this situation, the law permits a shareholder (the "derivative plaintiff") to sue the directors and officers in the name of, and on behalf of the corporation. The derivative plaintiff need merely demonstrate that a majority of the board of directors lack independence of judgment in this dispute, excusing a demand upon the directors and officers to sue themselves. The shareholder may be eligible to receive an incentive award, in the discretion of the court, as part of the settlement of a shareholder derivative action. An incentive award compensates you for your time and trouble in bringing the action, and rewards you for defending the rights of the corporation. Such awards often exceed the individual damages of the shareholder bringing the action. Notably, incentive awards are unavailable to plaintiffs in securities fraud class actions.

Whistle-blower Cases

If you are aware of a company that is defrauding the government, you may be eligible to bring a whistle-blower case under the False Claims Act. Defense contractor fraud, Medicaid fraud, Medicare fraud and Medical frauds are common examples of situations where companies bill the government for monies to which they are not entitled, costing the government billions of dollars. You may bring a False Claims Act case on behalf of the government and be entitled to an award of a portion of the recovery.

Securities Fraud Class Actions – Frequently Asked Questions

What is a class representative?

A class representative is a person who sues on behalf of a group of other shareholders and seeks to recover not only his own damages, but those of an entire class of defrauded shareholders. You must meet certain qualifications in order to serve as class representatives:

- You must not have purchased the security in order to participate in the litigation, or at the direction of your attorney.
- You must have actually suffered damages in the fraud (i.e., if you made money in the stock, even though a fraud occurred, you are not a victim);
- If you have served as a class representative in five cases in the last three years, you must seek leave of court to serve again;
- You must agree not to accept any payment for serving as class representative beyond your pro rata share of any recovery, except your reasonable costs and expenses (including lost wages) directly relating to the representation of the class; award of these amounts must be approved by the Court.

Does it matter if I already sold my shares?

No, you were defrauded when you bought the shares. However, it may affect the amount of your damages. If you sold your shares and you made money, you probably don't have a claim.

Do I need to hold my shares to participate?

No, you may sell your shares at any time. (This is distinct from a shareholder derivative suit, in which you must continue to hold at least some of your shares while the lawsuit is pending.)

Will this cost me any money?

No, once we have accepted you as a client, we will advance the costs of your case and seek reimbursement of costs and payment of our attorney's fee from the court only upon successful conclusion of your case. If we are not successful, you will not be responsible for costs or fees.

How long will this take?

A securities fraud class action is rarely settled in much less than a year. More commonly, such cases take from two to four years to resolve. If tried and appealed, the case can last six years or more.

What can I expect to recover?

You may expect to recover your pro rata share of your damages. Often the total funds available to pay damages in settlement, or after judgment, are inadequate to pay each shareholder in full. This is because the parties responsible for the fraud usually do not have enough assets or insurance coverage to make full payment. In either event, judgment or settlement, the total amount recovered is usually some fraction of the whole. You may expect to receive your pro rata portion of that amount.

How are legal fees calculated?

We keep track of our time spent on the case and maintain detailed records of all of our activities on the case and all of our expenses. When we apply for a fee at the end of the case, we may request a percentage of the recovery, but the court will usually want to see our billing records before it decides. In federal courts, fees are commonly awarded on a percentage basis, with the actual percentage awarded ranging from 20% up to 33% (if you have seen reports in the press of higher percentages, we believe them to be unfounded or simply exceptional). In California state court, fees are generally awarded based on "lodestar" (hours x hourly rate = lodestar), plus a risk factor multiplier. The court will assess the difficulty of the case a number of different ways and then decide if the lawyers deserve some multiple of their time in the case. Such a multiple can range from around three all the way down to a 'negative multiplier' (a fraction less than 1) if the court believes the requested fee to be excessive.

"Pennies for the shareholder, millions for the lawyers?"

Class action cases have generated a fair amount of controversy in the press in recent years. Critics claim that the typical class member recovers pennies, while the typical class lawyer recovers millions in fees. The implication is that you should not participate and try to recover your loss, because you won't get much back anyway, and besides, you'll just be making some lawyer rich. This makes our blood boil. Here's why:

The USA is one of the very few places in the world where the average citizen has access to the courts. In most countries, justice is for the rich. There are those in the US who would like to reduce access to the courts by the average citizen. Typically, these are corporate interests or accounting professionals that would just as soon avoid liability, any liability, for their products and services, no matter what harm they have caused. They are well-funded and spend their money in Congress and the state legislatures, perpetually seeking to reduce their collective exposure.

The class action device is quintessentially democratic. The theory behind it is that, while a single individual's claim may be too small to pursue in court, the same defendant may have wronged enough people in the same way so as to justify aggregating all the individual claims. The device answers the question: 'Can I get away by stealing $50 million, just so long as I steal it from 10 million people just $5 at a time?' The answer is "No! You will be sued in a class action." Securities fraud class action cases are high stakes, high risk contingent litigation. Damages in these cases can measure in the billions of dollars. They are fantastically expensive to prosecute, costing anywhere from hundreds of thousands to millions of dollars to take through the discovery process, trial and appeals. Every penny of that money comes from the lawyers, who must fund the case to its successful conclusion before they recover a nickel of their costs. It can take four years or longer to resolve a securities fraud class action. Many securities fraud class action cases are dismissed by the courts, often well into the case. Because the defendants commonly control all of the important documents and knowledge in the case, it can be very difficult to prove the fraud and prevail in the court.

The civic rationale of class actions

At root, the securities laws in the United States are designed to deter misconduct. Class action cases have emerged as one of the most powerful deterrents to fraud in our securities markets. If you commit fraud on your shareholders, you will probably not be prosecuted by the SEC (after all, its resources are limited), but you may be sure that you will be sued by your shareholders. And if the SEC does prosecute, while it may punish the wrongdoers, it almost never recovers your losses. Only a civil lawsuit, such as a securities fraud class action, can do that. By suing those who have defrauded you, even if your own damages are modest, and even if you don't recover every penny, you are helping to keep the market honest and make it a safer place to invest your money. By depriving wrongdoers of the proceeds of their frauds, you are deterring fraud and punishing those who commit fraud. The result of this system of civil deterrence is manifest: the US has the largest, safest, healthiest markets in the history of the world. Look abroad and all too often you will see turmoil in the markets, shoddy accounting practices, two sets of books, overvalued inventory and receivables that never quite turn into revenue. The securities fraud class action is your weapon to see that it doesn't happen here.

13

36 LAWS OF WRONGDOING TO MINORITY SHAREHOLDERS IN UNETHICAL COMPANIES

"Les vertus se perdent dans l'interet comme les fleuves se perdent dans la mer."
Virtues get lost by personal interests like rivers that disappear into the sea.
(La Rochefoucauld, Maximes)

1 – In unethical companies, the minority shareholders will always lose in the long run.

2 – Unethical managers tend to work on the verge of the law, finding loopholes, and getting the legal advice of the best lawyers, in cases of wrongdoing to the minority shareholders.

3 – Boards of Directors and executives of companies tend to safeguard primarily the interests of the majority or controlling shareholders, who have appointed and remunerate them.

4 – Independent Directors, who are appointed by the executives, decisions of their committees, and fairness opinions that they order, are in many cases unreliable to minority shareholders, as they tend to comply with the opinions of the majority shareholders.

5 – Auditors, underwriters, analysts, investment bankers, and consultants are loyal primarily to the executives who remunerate them, and the minority shareholders should be cautious with their reports and recommendations.

6 – When examining the reports of analysts and their 'buy' suggestions on companies, one should bear in mind what are the interests of the analysts, if they own shares of the companies, and what is their success record until now.

7 – The legal system does not safeguard in most of the cases the rights of the minority shareholders, who cannot fight on equal terms with the companies that are assisted by the best lawyers, and have much more time and resources.

8 – Companies tend sometimes to accommodate large institutions, which were wronged as minority shareholders, mainly by indirect compensation.

9 – The SEC is in many cases a panacea that is indifferent to wrongdoing to minority shareholders and to creative accounting.

10 – Society does not ostracize unethical managers and believes that ethics should be confined to the observance of the laws.

11 – Minority shareholders should refrain from investing in companies whose ultimate goal is to maximize profits, as it would in many cases benefit only the profits of the majority shareholders and executives.

12 – Minority shareholders should invest in companies having ethical CEOs, as they would probably safeguard their rights and not be loyal exclusively to the majority shareholders.

13 – Minority shareholders are often perceived as speculators, who do not care for the welfare of their companies, but are greedy and interested in an immediate and riskless return on investment.

14 – The perception of the minority shareholders as greedy and speculators, and the lack of personification to the nameless individuals, legitimize in many cases wrongdoing to them.

15 – Unethical companies tend to avoid transparency and publish opaque prospectuses, press releases and financial statements. Transparency is therefore the main safeguard of the minority shareholders.

16 – Shareholders should compare the prospectuses with the press releases and interviews of the executives and owners of the companies. If there is double talk and the information released to the SEC does not comply with the press conferences, it could indicate that the companies are in trouble.

17 – Minority shareholders should read carefully all the information accessible to them, participate in the stock talks on the Internet, and have a fair understanding of financial statements. If not, they should abstain from investing directly in companies and should rather invest in Ethical Funds.

18 – The conduct of the shares' price prior and subsequent to a public offering indicates the ethics of a company, especially if price increases substantially before the offering and collapses a short time afterwards.

19 – Minority shareholders should avoid investing in companies whose executives do not own their shares or have sold most of them, and whose controlling shareholders sell a large part of their shares at public offerings.

20 – Executives of many companies tend to receive warrants when the shares' price is at their lowest point and sell them at the end of their restriction period, when their prices reach a maximum. Minority shareholders are invited to read this information on the Internet and imitate their conduct.

21 – Unethical executives tend to benefit from inside information in buying and selling their shares and minority shareholders can receive indications on the future profitability of the company by following on the Internet insiders data. Selling of shares by insiders indicates future losses and buying of shares indicates favorable prospects.

22 – Companies that want to sell a subsidiary partially owned by them to a fully or majority owned affiliate company tend to convey the impression that the situation of the subsidiary is precarious, with no potential acquirers, in order to justify the collapse of its price and the acquisition of the subsidiary at a token price by the affiliate company.

23 – Unethical companies have double standards for their shareholders. They may convey the impression that they are on the verge of bankruptcy in order to discourage the minority shareholders, and after the controlling shareholders and executives buy their shares at minimal prices, make public encouraging prospects in order to increase their shares' price.

24 – Companies tend to be privatized before the end of revolutionary products' R&D or after the implementation of a successful turnaround plan, when the shares' prices are still low, by forcing the minority shareholders to sell their shares at those prices, and concealing those prospects to them.

25 – Delaware's Laws give extreme license to the controlling shareholders to do whatever they want in their companies and enable them in some cases to commit wrongdoing to minority shareholders without giving them a fair possibility of retaliation.

26 – Majority shareholders and executives tend to conceal their true motives of depriving the rights of the minority shareholders behind altruistic talks of saving employment, assisting the community and helping the economy.

27 – Minority shareholders should suspect government officials who are supposed to safeguard their rights if the law enables them to be recruited by the companies that they were supposed to control.

28 – Shares' transactions that are executed in August, during the vacations, around Christmas, New Year's eve, or in other periods, where most of the minority shareholders are out of town, are often meant to wrong them without giving them the opportunity to interfere.

29 – Shareholders' meetings are in many cases orchestrated in such a way that minority shareholders cannot express effectively their discontent, and even if they do so the protocols of the meetings do not report it.

30 – Minority shareholders should beware of companies that expense too often extraordinary losses, charges for in-process technology, acquisition costs, contingent liabilities, and make huge reserves for non-recurring charges on restructuring plans. Those losses may be a heaven, concealing operational losses, and precursory of the imminent collapse of the company's valuation.

31 – Minority shareholders should refrain from investing in companies that are controlled exclusively by the majority shareholders, especially if those own less than 50 percent of the shares, and allow no representation of the minority shareholders in their Boards of Directors.

32 – Activist associations should gather information on unethical companies, shareholders and executives and publish it on the Internet and to minority shareholders. People tend to forget or do not have access to this data and the activists' responsibility is to make the relevant information accessible to all.

33 – Disclosers of unethical conduct of companies toward minority shareholders should be encouraged by rewards, esteem and recognition, and should not be ostracized by society as whistle-blowers.

34 – Individual shareholders who have lost in the stock market, due to an unethical conduct of companies, should publish the information on the Internet, the press, the SEC, among their friends, and try to get the maximum coverage for the wrongdoing of unethical companies.

35 – Minority shareholders should only resort to ethical means if they have to fight the companies that have wronged them, as in an unethical combat the stronger parties will always win.

36 – The minority shareholders should put a very high emphasis on the ethics of the companies and the integrity of their managers and owners in their investing considerations and refrain from investing in unethical companies that might wrong them, even if those companies have excellent prospects.

14
CONCLUSION

"God loves from whole to parts; but human soul
Must rise from individual to the whole.
Self-love but serves the virtuous mind to wake,
As the small pebble stirs the peaceful lake;
The centre moved, a circle strait succeeds,
Another still, and still another spreads;
Friend, parent, neighbour, first will it embrace;
His country next; and next all human race."
(Alexander Pope, 1688-1744, An Essay on Man)

Toward the end of the book it is legitimate to ask: What is the gist of the book? Is this book an essay on poetry, ideals and altruism that should fulfill what Pope wrote three centuries ago about the human soul that should rise from individual to the whole? The drive to write the book came from a personal experience and from witnessing too many traumatic cases of wrongdoing to minority shareholders, flagrantly performed, without any fear of being punished by the existing panaceas. The new vehicles for the revolution are in their embryonic stage, the cases have proven how the activist associations, the Internet and transparency were not sufficient to win the cause of the wronged shareholders.

One and a half century have elapsed since the events that took place in the French stock exchange that inspired Zola to write his masterpiece L'Argent. His book that should be the bible of the minority shareholders concludes by describing the outcome of the schemes to which they have succumbed. Every small shareholder should read the following lines before deciding to invest in the stock exchange today as in the times of Zola. To them is this Business Ethics book dedicated with pity, compassion, and empathy.

"Mais les morts inconnus, les victimes sans nom, sans histoire, emplissaient surtout d'une pitie infinie le coeur de Mme. Caroline. Ceux-la etaient legion, jonchaient les buissons ecartes, les fosses pleins d'herbe, et il y avait ainsi des cadavres perdus, des blesses ralant d'angoisse, derriere chaque tronc d'arbre. Que d'effroyables drames muets, la cohue des petits rentiers pauvres, des petits actionnaires ayant mis toutes leurs economies dans une meme valeur, les concierges retires, les pales demoiselles vivant avec un chat, les retraites de province a l'existence reglee de maniaques, les pretres de campagne denudes par l'aumone, tous ces etres infimes dont le budget est de quelques sous, tant pour le lait, tant pour le pain, un budget si exact et si reduit, que deux sous de moins amenent des cataclysmes! Et, brusquement, plus rien, la vie coupee, emportee, de vieilles mains tremblantes, eperdues, tatonnantes dans les tenebres, incapables de travail, toutes ces existences humbles et tranquilles jetees d'un coup a l'epouvante du besoin!" (Emile Zola, L'Argent, p. 440)

"But the unknown dead, the nameless victims, with no history, filled especially with infinite pity the heart of Mme. Caroline. Those were legions, were strewn all over the remote bushes, the ditches full of grass, lost corpses, wounded people moaning from anxiety, behind every trunk of a tree. How many dreadful silent dramas, the crowd of the small poor retired people, the small shareholders who have invested all their savings in the same stock, the retired concierges, the pale old maids living with a cat, the old people living in the country in a well-ordered obsessive existence, the priests in the villages resorting to begging, all those tiny little people with tight budgets, so much for milk, so much for bread, such a small and exact budget, that any reduction can cause a cataclysm! And, all of a sudden, a void, life is cut off, taken away, old shaky hands, desperate, groping in the dark, unable to work, all those humble and quiet lives thrown all of a sudden to the terror of poverty!"

But closer to our times, only half a century ago, the prophetic poem of Pope still remained as remote as ever. The same ethical dilemmas remain and become even more acute. The monumental play of Arthur Miller, All My Sons, can be treated as a case study in business ethics as it summarizes the principal themes of the book. From each scene one can draw a conclusion that refers to one of the chapters of the book. This book has tried to juxtapose professional literature on ethics, classical literature with ethical subjects, and real life cases of ethical dilemmas in American, French and Israeli companies. The conclusion of the book will be faithful to this method, which emphasizes the reality of fiction.

Joe Keller is a rich American industrial, who has sent during the war defective aircraft parts to the Air Force, causing the death of 21 pilots. His son, Larry, a pilot himself, who hears those news during the war, disappears with his aircraft. By the end of this unforgettable play, we learn that before dying, Larry has sent to his fiancee Ann a farewell letter explaining to her that he can no longer live with a guilty conscience of his father's crime. Joe is an ethics criminal, but legally he has managed to be acquitted. He returns home after the acquittal, and society exculpates him.

"Everybody knew I was getting out that day; the porches were loaded. Picture it now; none of them believed I was innocent. The story was, I pulled a fast one getting myself exonerated. So I get out of my car, and I walk down the street. But very slow. And with a smile. The beast! I was the beast, the guy who sold cracked cylinder heads to the Army Air Force; the guy who made twenty-one P-40's crash in Australia. Kid, walkin' down the street that day I was guilty as hell. Except, I wasn't, and there was a court paper in my pocket to prove I wasn't, and I walked... past... the porches. Result? Fourteen months later I had one of the best shops in the state again, a respected man again; bigger than ever." (Six Great Modern Plays, Arthur Miller, All My Sons, p. 381-2)

His environment forgives him, as he is rich, has a lot of nerve, is self-confident, he managed to outsmart the court, and who cares if he has caused the death of 21 pilots... Society is therefore an accessory to Joe Keller's crime, with its benevolence to ethics criminals. Conclusion, as long as society will let criminals get away with their ethical crimes and will not ostracize them, it will be very difficult to fight effectively ethical crimes, as the law will almost always exculpate the criminals, who are often the strongest and smartest, and have at their disposal the best lawyers and the largest funds.

Joe Keller, is worried that Ann intends to marry his son Chris, that Ann's father who was his partner and his crime accomplice is going to be released from prison, and that Ann's brother, George Deever, suspects his complicity. Joe tries to corrupt them by offering a job in his factory to George and to his father, they try to marry George with a friend, they embrace him with attention. When it does not succeed, Joe threatens George and blames his father for the crime he himself has committed. Conclusion, all is permitted to safeguard your interests - corruption, threats, deceit. You blame the others for your own crimes. The victims are the 'speculators' of Joe Keller, the lambs are treated as wolves, and the wolves are disguised as innocent grandmothers. From the moment we start to behave unethically, there are no more limits.

But one should nevertheless keep up appearances, even at the price of self deceit. Kate Keller, Joe's wife, who knows perfectly well that Joe is guilty, refuses to admit that Larry is dead, because if she admits it, it would mean that Joe had murdered his own son. She is therefore forced to oppose the wedding of Chris with Ann, as Larry has to remain alive and disappeared. It is her compromise with her conscience, but it is exactly this lie that is the basis of the denouement, as it is impossible to base your existence on lies. Her conscience allows her to admit that her husband has murdered 21 'anonymous' pilots, but does not let her admit that he murdered his own son, as if there was a difference between blood and blood. "Your brother's alive, darling, because if he's dead your father killed him. Do you understand me now? As long as you live, that boy is alive. God does not let a son be killed by his father." (Miller, All My Sons, p. 418)

Conclusion, one can always compromise with his conscience on all degrees of ethical and other crimes. Joe is convinced that he did not murdered the pilots, as he does not know them personally, he does not personify them, exactly like in Marcel Pagnol's papet. We can commit a crime against Jean de Florette as he is a stranger, we can kill 21 pilots, we can wrong the rights of minority shareholders, as we do not know them, they are weak and cannot retaliate.

Chris, Joe's son who works with his father, suspects the culpability of his father. He is the member of the second generation of ethics criminals, a precursor of Michael, the son of Don Corleone, who has started his career with good intentions but who was ultimately corrupted by his environment. In 1947, the year Miller wrote his masterpiece, as in the year 2000, the ethical dilemma is the same. Chris appeases his conscience by saying that everything is permitted in the business world: "This is the land of the great big dogs, you don't love a man here, you eat him! That's the principle; the only one we live by – it just happened to kill a few people this time, that's all. The world's that way, how can I take it out on him? What sense does that make? This is a zoo, a zoo!" (Miller, All My Sons, p. 429) But even Chris has his scruples and he cannot solve the dilemma between his conscience and the love and respect he owes to his father.

And as Joe does not succeed in convincing his son Chris, he tries the well-known stratagem, by telling him that 'everybody does the same', therefore - vox populi vox dei, a doctrine that led to the most violent crimes in history, as you do not measure your conduct according to your conscience or ethics, but according to what you perceive or you fool yourself to believe are the norms of society: "Who worked for nothin' in that war? When they work for nothin', I'll work for nothin'. Did they ship a gun or a truck outa Detroit before they got their price? Is that clean? It's dollars and cents, nickels and dimes; war and peace, it's nickels and dimes, what's clean? Half the Goddam country is gotta go if I go!" (Miller, All My Sons, p. 430) And everybody continues to join the parade…

Conclusion, everything is allowed if it is the norm of society even if it is unjust, as you cannot survive otherwise in the business world. Everything consists in dollars, francs or shekels, there is no other thing, war, peace, morals, ethics, family, friends, country. From the moment that we admit that everything is based on profits, everything is really permitted. It is therefore the complete bankruptcy of the business world, an obscure world, dangerous, merciless, without compassion, where everybody is a wolf to the other, where everything is allowed as long as you earn nickels and dimes.

At the end of the play, Joe learns that his son Larry has not disappeared but has committed suicide, without being able to forgive his father on his crime and assassination of 21 pilots, and his last words in his farewell letter to Ann are - that if he could he would have killed his own father for what he did. Joe is at last affected by his crime, which is personalized by the curse of his son. He quits his wife and his life, and before committing suicide he says: "Sure, he was my son. But I think to him they were all my sons. And I guess they were, I guess they were." (Miller, All My Sons, p. 432)

Conclusion, from the moment that you admit that all are your sons and brothers, that you should not do unto the other what you would not want to be done to you, we return to the biblical, philosophical and literary precepts of love of others, which should be at the basis of the business world, and destroy the last vestiges of autocracy and cannibalism, of belligerence and corruption. In a world where 'cut throat competition' is a leit motive, where 'street fighter' is a hero, where the CEO is king, and where there are no scruples, one always murders ultimately his son and one has to commit suicide, at least virtually, by suppressing his conscience and killing it completely.

And the moral of the play and the book, is in the last words of Chris: "You can be better! Once and for all you can know there's a universe of people outside and you're responsible to it, and unless you know that, you threw away your son because that's why he died." (Miller, All My Sons, p. 432) The conclusion in our context is that beyond the company, the board of directors, the executives and the majority shareholders, there are also the stakeholders, the employees, the suppliers, the customers, the community, the nation, and also the minority shareholders. From the moment you forget them, when you only see your own portrait in the silver mirror, when you disregard the world, the environment and others, when you are no longer responsible toward your brothers and you are looking only after you own interests, it is equivalent to the murder of your own sons.

And those of us who do not want to murder their sons, their conscience, their neighbors, have to resort to ethics that will guide them to a happy life, as maintained by Aristotle, a life of moderation, psychological, spiritual, and emotional equilibrium, that maybe will not give them the maximum profitability but will ensure them a successful combination of profitability with a peace of mind. The majority, the privileged, the people who control society have already reached the conclusion that the best way to govern a country is through democracy, where every minority, the weak and the underprivileged have also a fair share of the country governance.

This book has suggested many practical methods and vehicles to obtain democracy in the business world as well. It can be done in the hard way by struggle and fight as happened in the democratic process in France and Russia, but it can be achieved in a subtler way by understanding and concord as was achieved in Great Britain and Scandinavia.

The author of the book is convinced that the subtler way is the best and that the weaker parties in the economy, including the minority shareholders, have to do their utmost to reach an understanding with the people who control the economy before resorting to the Armageddon weapon of breaking the system and reconstructing it on a better basis.

But time is of the essence as it is running out, and the people who rule the economy and the companies have also to reach the conclusion that the best way to control the companies is by sharing the power with the weaker parties, who could be the stronger if they organize and exert their power.

The minority shareholders are in the middle of their long Odyssey. There is a long way still to Ithaca, Poseidon is still winning and Athena cannot rescue them. They have encountered in their long journey Cyclops, Lotus-eaters, the monsters Scylla and Chariybdis, and even cannibalistic Laistrygones, who ate in many cases all their savings voraciously. The small shareholders ate the lotus, forgot what happened to them and were unmindful of the wrongdoing. They were lured by the Sirens, the enchantress Circe and Calypso, who enticed them to invest in many dubious schemes. They have even visited the land of the Departed Spirits, where they have met all the millions of their predecessors who over the hundreds of years that have elapsed since the first shareholders' scandals have lost all their money and plunged into poverty. They are still marching in the endless caravan, waiting for Ulysses to lead them.

Thousands of years have elapsed since the time of Homer and the time has come to deliver the minority shareholders from their servitude. Many leaders have succeeded in managing the campaigns of the customers, quality and the environment. The leader of the minority shareholders' campaign will have to enable the Internet, the Disclosers and Transparency to act effectively as Trojan Horses, which will destroy the citadels of the unethical companies that were until now impregnable. He will be assisted by the activist associations, which are still in many cases weak and inefficient, but are getting momentum. In the meantime, the minority shareholders should invest in Ethical Funds and refrain from investing in the Stock Exchange in unethical companies.

And ultimately, the stakeholders and the minority shareholders will get their due rights and cease to be exploited and wronged, when every one of us will understand that we are all stakeholders, customers, suppliers or members of the community, all of us are minority shareholders, and everyone has to believe, act and say to himself: 'For Me, You Are All My Sons!'

BIBLIOGRAPHY

Aristotle, Ethics, Penguin Classics, 1976

Badaracco, Jr. Joseph L., Defining Moments, When Managers Must Choose between Right and Right, Harvard Business School Press,1997

The Holy Bible, The Old Testament, New International Version, International Bible Society, 2/96

The Holy Bible, The New Testament, New International Version, International Bible Society, 2/96

Blanchard Ken, O'Connor Michael, Managing By Values, Berrett-Koehler Publishers, 1997

Blanchard Kenneth, Peale Norman Vincent, The Power of Ethical Management, William Morrow and Company, Inc., 1988

Blaug Mark, La Methodologie Economique, The Economic Methodology, in French, 2e edition, Ed. Economica 1994

Bollier David, Aiming Higher, 25 Stories of how Companies Prosper by Combining Sound Management and Social Vision, Amacom, 1997

Bonder Nilton Rabbi, The Kabbalah of Money, Insights on Livelihood, Business, and All Forms of Economic Behavior, Shambhala, 1996

Brecht Bertolt, Die Dreigroschenoper, The Threepenny Opera, in German, Universal Edition, 1928, assigned to Brookhouse Music, Inc., 1957

Briner Bob, The Management Methods of Jesus, Ancient Wisdom for Modern Business, Thomas Nelson, Inc., 1996

Burkett Larry, Business by the Book, The Complete Guide of Biblical Principles for Business Men and Women, Expanded Edition, Thomas Nelson Publishers, 1991

Burkett Larry, Sound Business Principles, Includes Ethics and Priorities, Moody Press, 1993

Business Ethics Quarterly, January 1998, Vol. 8 No. 1, The Journal of the Society for Business Ethics

Business Ethics Quarterly, April 1998, Vol. 8 No. 2, The Journal of the Society for Business Ethics

Business Ethics Quarterly, July 1998, Vol. 8 No. 3, The Journal of the Society for Business Ethics

Business Ethics Quarterly, October 1998, Vol. 8 No. 4, The Journal of the Society for Business Ethics

Business Ethics Quarterly, The Ruffin Series Special Issue No. 1, 1998, The Journal of the Society for Business Ethics

Business Ethics Quarterly, January 1999, Vol. 9 No. 1, The Journal of the Society for Business Ethics

Business Ethics Quarterly, April 1999, Vol. 9 No. 2, The Journal of the Society for Business Ethics

Business Ethics Quarterly, July 1999, Vol. 9 No. 3, The Journal of the Society for Business Ethics

Business Ethics Quarterly, October 1999, Vol. 9 No. 4, The Journal of the Society for Business Ethics

Business Ethics Quarterly, The Ruffin Series Special Issue No. 2, 2000, Environmental Challenges to Business, A Publication of the Society for Business Ethics

Business Ethics Quarterly, January 2000, Vol. 10 No. 1, The Journal of the Society for Business Ethics

Casey Al with Seaver Dick, Casey's Law, If Something Can Go Right, It Should, Arcade Publishing, 1997

Cervantes Miguel de, El Ingenioso Hidalgo Don Quijote de la Mancha, Don Quixote, in Spanish, Catedra Letras Hispanicas, 1992

Chatfield Cheryl A., Ph.D., The Trust Factor, The Art of Doing Business in the 21st Century, Sunstone Press, 1997

Cohen Ben and Greenfield Jerry, Ben & Jerry's Double-Dip, Lead with Your Values and Make Money, Too, Simon & Schuster, 1997

Dante, The Divine Comedy, Pan Classics, 1980

De George Richard T., Competing with Integrity in International Business, Oxford University Press, 1993

Derber Charles, The Wilding of America, How Greed and Violence Are Eroding Our Nation's Character, St. Martin's Press, 1996

Devine George, Responses to 101 Questions on Business Ethics, Paulist Press, 1996

Dherse Jean-Loup, Minguet Hughes Dom, l'Ethique ou le Chaos?, Ethics or Chaos?, in French, Presses de la Renaissance, 1998

Donaldson Thomas, The Ethics of International Business, The Ruffin Series in Business Ethics, Oxford University Press, 1992

Driscoll Dawn-Marie, Hoffman W. Michael, Petry Edward S., The Ethical Edge, Tales of Organizations that Have Faced Moral Crises, MasterMedia Limited, 1995

Estes Ralph, Tyranny of the Bottom Line, Why Corporations Make Good People Do Bad Things, Berrett-Koehler Publishers, 1996

Ethics at Work, A Harvard Business Review Paperback, 1991

Etzioni Amitai, The Moral Dimension, Toward a New Economics, The Free Press, 1990

Feyerabend Paul, Contre la Methode, Esquisse d'une Theorie Anarchiste de la Connaissance, Against Method, in French, Editions du Seuil, 1979

Fukuyama Francis, Trust, The Social Virtues and the Creation of Prosperity, A Free Press Paperbacks Book, 1996

Ginsburg Sigmund G., Managing with Passion, Making the Most of Your Job and Your Life, John Wiley & Sons, Inc., 1996

Hall William D., Making the Right Decision, Ethics for Managers, John Wiley & Sons, Inc., 1993

Handy Charles, The Hungry Spirit, Beyond Capitalism: A Quest for Purpose in the Modern World, Broadway Books, 1998

Harmon Frederick G., Playing for Keeps, How the World's Most Aggressive and Admired Companies Use Core Values to Manage, Energize and Organize Their People, and Promote, Advance and Achieve Their Corporate Missions, John Wiley & Sons, Inc., 1996

Harvey Brian, Edited by, Business Ethics, A European Approach, Prentice Hall, 1994

Homer, The Iliad, Penguin Classics, 1987

Homere, Odyssee, Le Livre de Poche, 1960

Hornstein Harvey A., Ph.D., Brutal Bosses and Their Prey, Riverhead Books, 1996

Huckabee Mike, the Honorable Governor of Arkansas, with Perry John, Character is the Issue, How People with Integrity Can Revolutionize America, Broadman & Holman Publishers, 1997

Ibsen Henrik, An Enemy of the People, Modern Library College Editions, 1950

Inoue Shinichi, Putting Buddhism to Work, A New Approach to Management and Business, Kodansha International Ltd., 1997

Jackall Robert, Moral Mazes, The World of Corporate Managers, Oxford University Press, 1989

Jackson Jennifer, An Introduction to Business Ethics, Blackwell Publishers, 1996

Jacobs Joseph J., The Anatomy of an Entrepreneur, Family, Culture, and Ethics, ICS Press, 1991

Jay Antony, Management and Machiavelli, in Hebrew, Ma'ariv Book Guild, 1989

Kafka Franz, The Trial, Schocken Books, 1998

Kafka Franz, The Metamorphosis and Other Stories, Barnes an Noble Books, 1996

Kaufman Allen, Zacharias Lawrence, Karson Marvin, Managers vs. Owners, The Struggle for Corporate Control in American Democracy, The Ruffin Series in Business Ethics, Oxford University Press, 1995

Kelley Michael, On Stone or Sand, The Ethics of Christianity, Capitalism, & Socialism, Pleroma Press, 1993

Kidder Rushworth M., How Good People Make Tough Choices, Resolving the Dilemmas of Ethical Living, A Fireside Book published by Simon & Schuster, 1995

Koran, the Essential, the Heart of Islam, an Introductory Selection of Readings from the Qur'an, Translated and Presented by Thomas Cleary, Castle Books, 1993

La Fontaine, Fables, in French, Folio, Gallimard, 1991

Lawrence William D., with Turpin Jack A., Beyond the Bottom Line, Where Faith and Business Meet, Praxis Books Moody Press, 1994

Leibowitz Yeshayahu, Talks with Michael Shashar, On Just About Everything, in Hebrew, Keter Publishing House, 1988

Lynn Jonathan and Jay Antony, edited by, The Complete Yes Minister, The Diaries of a Cabinet Minister by the Right Hon. James Hacker MP, British Broadcasting Corporation, 1985

Machiavelli Niccolo, The Prince, Bantan Books, 1981

Madsen Peter, Ph.D., and Shafrtiz Jay M., Ph.D., Essentials of Business Ethics, A Collection of Articles by Top Social Thinkers, Including Peter Drucker, Milton Friedman, Robert Jackall, Ralph Nader, Laura Nash, Patricia H. Werhane, A Meridian Book, 1990

Mao Tse-Tung, Quotations from Chairman, Foreign Languages Press, 1966

May William W., Business Ethics and the Law, Beyond Compliance, The Rockwell Lecture Series, Peter Lang, 1991

Miller Arthur, All My Sons, Six Great Modern Plays, Dell Publishing Company, Inc., 1977

Miller Arthur, The Crucible, The Portable Arthur Miller, Penguin Books, 1995

Monks Robert A.G., The Emperor's Nightingale, Restoring the Integrity of the Corporation in the Age of Shareholder Activism, Addison-Wesley, 1998

Mott Graham M., How to Recognize and Avoid Scams Swindles and Rip-Offs, Personal Stories Powerful Lessons, Golden Shadows Press, 1994

Nash Laura L., Ph.D., Believers in Business, Resolving the Tensions between Christian Faith, Business, Ethics, Competition and our Definitions of Success, Thomas Nelson Publishers, 1994

Nash Laura L., Good Intentions Aside, A Manager's Guide to Resolving Ethical Problems, Harvard Business School Press, 1993

O'Neill Jessie H., The Golden Ghetto, The Psychology of Affluence, Hazelden, 1997

Pagnol Marcel, Judas, in French, Oeuvres Completes I, Theatre, Editions de Fallois, 1995

Pagnol Marcel, Les Marchands de Gloire, The Merchants of Glory, in French, Oeuvres Completes I, Theatre, Editions de Fallois, 1995

Pagnol Marcel, Topaze, in French, Oeuvres Completes I, Theatre, Editions de Fallois, 1995

Pagnol Marcel, L'Eau des Collines, Jean de Florette, in French, Oeuvres Completes III, Souvenirs et Romans, Editions de Fallois, 1995

Pagnol Marcel, L'Eau des Collines, Manon des Sources, in French, Oeuvres Completes III, Souvenirs et Romans, Editions de Fallois, 1995

Parks Robert H., Ph.D., The Witch Doctor of Wall Street, A Noted Financial Expert Guides You through Today's Voodoo Economics, Prometheus Books, 1996

Passeron Jean-Claude, Le Raisonnement Sociologique, L'Espace Non-Popperien du Raisonnement Naturel, Sociological Reasoning, in French, Collection Essais & Recherches, Nathan, 1991

Peters Thomas J. and Waterman, Jr. Robert H., In Search of Excellence, Lessons from America's Best-Run Companies, Warner Books, 1984

Phillips Michael and Rasberry Sally, Honest Business, A Superior Strategy for Starting and Managing Your Own Business, Shambhala, 1996

Piave Francesco Maria, Rigoletto, in Italian, TMK(S), Marca Registrada RCA Corporation, 1974

Popper Karl R., La Logique de la Decouverte Scientifique, The Logic of Scientific Discovery, in French, Bibliotheque Scientifique Payot, Editions Payot, 1995

Pratley Peter, The Essence of Business Ethics, Prentice Hall, 1995

Quintus Smyrnaeus, The Fall of Troy, Loeb Classical Library, 1984

Rae Scott B. & Wong Kenman L., Beyond Integrity, A Judeo-Christian Approach to Business Ethics, Zondervan Publishing House, 1996

Richardson Janice, edited by, World Ethics Report on Finance and Money, Editions Eska, 1997

Scott Fitzgerald Francis, The Great Gatsby, Heinemann/Octopus, 1977

Shakespeare William, Julius Caesar, Oxford University Press, 1959

Shakespeare William, The Merchant of Venice, Oxford University Press, 1959

Solomon Robert C., Above the Bottom Line, An Introduction to Business Ethics, Second Edition, Harcourt Brace College Publishers, 1994

Solomon Robert C., Ethics and Excellence, Cooperation and Integrity in Business, The Ruffin Series in Business Ethics, Oxford University Press, 1993

Strauss Anselm L., Qualitative Analysis for Social Scientists, Cambridge University Press, 1996

248

Strauss Michael, Volition and Valuation, in Hebrew, Haifa University Press & Zmora-Bitan, Publishers, 1998

Tamari Meir, The Challenge of Wealth, A Jewish Perspective on Earning and Spending Money, Jason Aronson Inc., 1995

Troyat Henri, Zola, Flammarion, in French, Le Livre de Poche, 1992.

Velasquez Manuel G., Business Ethics, Concepts and Cases, Fourth Edition, Prentice Hall, 1998

Voltaire, Candide, Dover Publications, Inc., 1993

Wallwork Ernest, Psychoanalysis and Ethics, Yale University Press, 1991

Ward Gary, Developing & Enforcing a Code of Business Ethics, A Guide to Developing, Implementing, Enforcing and Evaluating an Effective Ethics Program, Pilot Books, 1989

Williams Oliver F., Houck John W., edited by, A Virtuous Life in Business, Stories of Courage and Integrity in the Corporate World, Rowman & Littlefield Publishers, Inc., 1992

Williams Oliver F., Reilly Frank K. & Houck John W., edited by, Ethics and the Investment Industry, Rowman & Littlefield Publishers, Inc., 1989

Wilson Rodney, Economics, Ethics and Religion, Jewish, Christian and Muslim Economic Thought, New York University Press, 1997

Woodstock Theological Center, Seminar in Business Ethics, Ethical Considerations in Corporate Takeovers, Georgetown University Press, 1990

Wright Lesley and Smye Marti, Corporate Abuse, How "Lean and Mean" Robs People and Profits, Macmillan, 1996

Wuthnow Robert, Poor Richard's Principle, Recovering the American Dream through the Moral Dimension of Work, Busines, & Money, Princeton University Press, 1996

Yin Robert K., Case Study Research, Design and Methods, Second Edition, Applied Social Research Methods Series, Volume 5, Sage Publications, 1994

Zola Emile, La Curee, The Quarry, in French, Gallimard, 1997

Zola Emile, L'Argent, Money, in French, Fasquelle, Le Livre de Poche 584, 1992

Zola Emile, Le Ventre de Paris, The Stomach of Paris, in French, Gallimard, 1996

LIST OF ARTICLES

Agle Bradley R. and Van Burren III Harry J., God and Mammon: The Modern Relationship, Business Ethics Quarterly, 1999 (4)

Amiel Barbara, Feminist Harassment, National Review, Rae, Beyond Integrity

Andreas Kurt, Germans and the D-Mark, Richardson, World Ethics Report on Finance and Money

Andrews Kenneth R., Ethics in Practice, Ethics at Work, A Harvard Business Review Paperback, 1991

Argandona Antonio, Business, law and regulation: ethical issues, Harvey, Business Ethics: a European Approach

Arrington Robert L., Advertising and Behavior Control, Journal of Business Ethics, Rae, Beyond Integrity

Auerbach Joseph, The Poletown Dilemma, Ethics at Work, A Harvard Business Review Paperback, 1991

Bandow Doug, Environmentalism: The Triumph of Politics, The Freeman, Rae, Beyond Integrity

Barry Vincent, Advertising and Corporate Ethics, Madsen, Essentials of Business Ethics

Bartolome Fernando, Nobody Trusts the Boss Completely – Now What?, Ethics at Work, A Harvard Business Review Paperback, 1991

Bass Kenneth, Barnett Tim, and Brown Gene, Individual Difference Variables, Ethical Judgments, and Ethical Behavior Intentions, Business Ethics Quarterly, 1999 (2)

Batakovic Dusan T., To Obey Is to Survive, Richardson, World Ethics Report on Finance and Money

Bazerman Max H. and Messick David M., On the Power of a Clear Definition of Rationality, Business Ethics Quarterly, 1998 (3)

Betz Joseph, Business Ethics and Politics, Business Ethics Quarterly, 1998 (4)

Bhide Amar and Stevenson Howard, Why Be Honest If Honesty Doesn't Pay, Ethics at Work, A Harvard Business Review Paperback, 1991

Bicchieri Cristina and Fukui Yoshitaka, The Great Illusion: Ignorance, Informational Cascades, and the Persistence of Unpopular Norms, Business Ethics Quarterly, 1999 (1)

Binmore Ken, Game Theory and Business Ethics, Business Ethics Quarterly, 1999 (1)

Boatright John R., Does Business Ethics Rest on a Mistake? Business Ethics Quarterly, 1999 (4)

250

Boatright John R., Globalization and the Ethics of Business, Business Ethics Quarterly, 2000 (1)

Bok Sissela, Whistleblowing and Professional Responsibility, New York University Education Quarterly, Rae, Beyond Integrity

Bouckaert Luk, Business and community, Harvey, Business Ethics: a European Approach

Bowie Norman E., Business Ethics and Cultural Relativism, Madsen, Essentials of Business Ethics

Bowie Norman E., Does It Pay to Bluff in Business, Business Ethics, Rae, Beyond Integrity

Brenkert George G., Marketing and the Vulnerable, Business Ethics Quarterly, Special Issue #1, 1998

Brenkert George G., Marketing to Inner-City Blacks: PowerMaster and Moral Responsibility, Business Ethics Quarterly, 1998 (1)

Brenkert George G., Trust, Business and Business Ethics: An Introduction, Business Ethics Quarterly, 1998 (2)

Brenkert George G., Trust, Morality and International Business, Business Ethics Quarterly, 1998 (2)

Brock Gillian, Are Corporations Morally Defensible?, Business Ethics Quarterly, 1998 (4)

Brockway George P., The Future of Business Ethics, Williams, Ethics and the Investment Industry

Buchholz Rogene A., The Evolution of Corporate Social Responsibility, Madsen, Essentials of Business Ethics

Cabot Stephen J., Plant Closing Bill Will Give Many Employees Their Day in Court, Madsen, Essentials of Business Ethics

Cadbury Adrian Sir, Ethical Managers Make Their Own Rules, Ethics at Work, A Harvard Business Review Paperback, 1991

Camdessus Michel, The Financial Crisis in Mexico: Origins, Response from the IMF and lessons to Be Learnt, Richardson, World Ethics Report on Finance and Money

Carr Albert Z., Is Business Bluffing Ethical?, Madsen, Essentials of Business Ethics

Chamberlain Neil W., Corporations and the Physical Environment, Madsen, Essentials of Business Ethics

Child James W. and Marcoux Alexei M., Freeman and Evan: Stakeholder Theory in the Original Position, Business Ethics Quarterly, 1999 (2)

Ciminello Romeo, Banks and Ethical Funds, Richardson, World Ethics Report on Finance and Money

Ciulla Joanne B., Imagination, Fantasy, Wishful Thinking and Truth, Business Ethics Quarterly, Special Issue #1, 1998

Ciulla Joanne B., On Getting to the Future First, Business Ethics Quarterly, 2000 (1)

Collier Jane, Theorising the Ethical Organization, Business Ethics Quarterly, 1998 (4)

COB, International Harmonisation of Accounting Practice, Richardson, World Ethics Report on Finance and Money

Cooke Robert Allan and Young Earl, The Ethical Side of Takeovers and Mergers, Madsen, Essentials of Business Ethics

Corbetta Guido, Shareholders, Harvey, Business Ethics: a European Approach

Danley John, Beyond Managerialism, Business Ethics Quarterly, Special Issue #1, 1998

Davis Philip E., Why Might Institutional Investors Destabilise Financial Markets?, Richardson, World Ethics Report on Finance and Money

De George Richard T., Ethics and the Financial Community: An Overview, Williams, Ethics and the Investment Industry

De George Richard T., Business Ethics and the Challenge of the Information Age, Business Ethics Quarterly, 2000 (1)

Delhommais Pierre-Antoine, Banks – More to Gain than to Lose with the Single Currency, Richardson, World Ethics Report on Finance and Money

Del Ponte Carla, The Fight Against Money Laundering in Switzerland, Richardson, World Ethics Report on Finance and Money

DeMott Benjamin, Reading Fiction to the Bottom Line, Ethics at Work, A Harvard Business Review Paperback, 1991

Des Jardins Joseph R., Privacy in Employment, Moral Rights in the Workplace, Rae, Beyond Integrity

Dodds Susan et alia, Sexual Harassment, Social Theory and Practice, Rae, Beyond Integrity

Donaldson Thomas, Multinational Decision-Making: Reconciling International Norms, Journal of Business Ethics, Rae, Beyond Integrity

Donaldson Thomas, Are Business Managers "Professionals"?, Business Ethics Quarterly, 2000 (1)

Drucker Peter, The Ethics of Responsibility, Madsen, Essentials of Business Ethics

Dunfee Thomas W., The Marketplace of Morality: Small Steps Toward a Theory of Moral Choice, Business Ethics Quarterly, 1998 (1)

Duska Ronald, Business Ethics: Oxymoron or Good Business?, Business Ethics Quarterly, 2000 (1)

Dwyer Paula, Shareholder Revolt, Richardson, World Ethics Report on Finance and Money

Eliet Guillaume, The Three Founding Principles of the Single European Stock Market, Richardson, World Ethics Report on Finance and Money

Endreo Gilles, Protection of Minority Shareholders in France, Richardson, World Ethics Report on Finance and Money

Estola Matti, About the Ethics of Business Competition, Business and Leadership Ethics, June 1998

Etzioni Amitai, A Communitarian Note on Stakeholder Theory, Business Ethics Quarterly, 1998 (4)

Ewing David W., Case of the Disputed Dismissal, Ethics at Work, A Harvard Business Review Paperback, 1991

Fadiman Jeffrey A., A Traveler's Guide to Gifts and Bribes, Ethics at Work, A Harvard Business Review Paperback, 1991

Faugerolas Laurent, Assessment of Stock Options in 1995, Richardson, World Ethics Report on Finance and Money

Ferrell O. C. and Fraedrich John, Understanding Pressures That Cause Unethical Behavior in Business, Business Insights, Rae, Beyond Integrity

Flores Fernando and Solomon Robert C., Creating Trust, Business Ethics Quarterly, 1998 (2)

Foegen J. H., The Double Jeopardy of Sexual Harassment, Business and Society Review, Rae, Beyond Integrity

Frederick William C., One Voice? Or Many? A Response to Ellen Klein, Business Ethics Quarterly, 1998 (3)

Freeman Edward R., Poverty and the Politics of Capitalism, Business Ethics Quarterly, Special Issue #1, 1998

Friedman Milton, The Social Responsibility of Business Is to Increase Its Profits, Madsen, Essentials of Business Ethics

Gaillard Jean-Michel, Retirement Management and Social Responsibility, Richardson, World Ethics Report on Finance and Money

Garaventa Eugene, Drama: A Tool for Teaching Business Ethics, Business Ethics Quarterly, 1998 (3)

Gauthier Frederic, A Summary of the Banking Crises in Central Europe, Latin America and Africa, Richardson, World Ethics Report on Finance and Money

Geisler Norman L., Natural Law and Business Ethics, Biblical Principles in Business: The Foundations, Rae, Beyond Integrity

Gellerman Saul W., Why "Good" Managers Make Bad Ethical Choices, Ethics at Work, A Harvard Business Review Paperback, 1991

Gerwen van Jef, Employers' and employees' rights and duties, Harvey, Business Ethics: a European Approach

Geva Aviva, Moral Problems of Employing Foreign Workers, Business Ethics Quarterly, 1999 (3)

Gibson Kevin, Bottom William, and Murnighan Keith J., Once Bitten: Defection and Reconciliation in a Cooperative Enterprise, Business Ethics Quarterly, 1999 (1)

Gini A. R. and Sullivan T., Work: The Process and the Person, Journal of Business Ethics, Rae, Beyond Integrity

Goodpaster Kenneth E., Business Ethics and Stakeholder Analysis, Business Ethics Quarterly, Rae, Beyond Integrity

Graddy Kathryn and Robertson Diana C., Fairness of Pricing Decision, Business Ethics Quarterly, 1999 (2)

Gross Joseph Prof., From the Desk of the Board of Directors – The New Corporate Law, Directors and Officers, Taxation Issues, in Hebrew, Globes, Israel, July 1999

Gross Joseph Prof., From the Desk of the Board of Directors, in Hebrew, October 1998

Gross Joseph Prof., From the Desk of the Board of Directors, in Hebrew, June 1998

Gross Joseph Prof., From the Desk of the Board of Directors, in Hebrew, March 1998

Gross Joseph Prof., From the Desk of the Board of Directors, in Hebrew, May 1997

Hamilton Stewart, How Safe Is Your Company?, Richardson, World Ethics Report on Finance and Money

Hanke Steve H., Argentina and the Tequila Effect, Richardson, World Ethics Report on Finance and Money

Hanke Steve H., Currency Board for Mexico, Richardson, World Ethics Report on Finance and Money

Hanson Kirk O., A Cautionary Assessment of Wall Street, Williams, Ethics and the Investment Industry

Hartman Edwin M., Altruism, Ingroups and Fairness: Comments on Messick, Business Ethics Quarterly, Special Issue #1, 1998

Hartman Edwin M., The Role of Character in Business Ethics, Business Ethics Quarterly, 1998 (3)

Hasnas John, The Normative Theories of Business Ethics: A Guide for the Perplexed, Business Ethics Quarterly, 1998 (1)

Hendry John, Universalizability and Reciprocity in International Business Ethics, Business Ethics Quarterly, 1999 (3)

Hoffman Michael W., Business and Environmental Ethics, Business Ethics Quarterly, Rae, Beyond Integrity

Hosmer Larue T., Lessons from the Wreck of the Exxon Valdez: The Need for Imagination, Empathy and Courage, Business Ethics Quarterly, Special Issue #1, 1998

Howard Robert, Values Make the Company: An Interview with Robert Haas, Ethics at Work, A Harvard Business Review Paperback, 1991

Husted Bryan W., The Ethical Limits of Trust in Business Relations, Business Ethics Quarterly, 1998 (2)

IMF Bulletin, How to Manage Today's Risks, Richardson, World Ethics Report on Finance and Money

Ishii Hiroshi, A Solution for the Crisis in the Japanese Banking System, Richardson, World Ethics Report on Finance and Money

Jackall Robert, Business as a Social and Moral Terrain, Madsen, Essentials of Business Ethics

Jackall Robert, Moral Mazes: Bureaucracy and Managerial Work, Ethics at Work, A Harvard Business Review Paperback, 1991

James Gene G., Whistle-Blowing: Its Moral Justification, Madsen, Essentials of Business Ethics

Jarrell Gregg A., The Insider Trading Scandal: Understanding the Problem, Williams, Ethics and the Investment Industry

Jensen Michael C., Takeovers: Folklore and Science, Harvard Business Review, Rae, Beyond Integrity

Jones Thomas M. and Verstegen Ryan Lori, The Effect of Organizational Forces on Individual Morality: Judgment, Moral Approbation, and Behavior, Business Ethics Quarterly, 1998 (3)

Kapstein Ethan B., Shockproof: the End of the Financial Crisis, Richardson, World Ethics Report on Finance and Money

Keller G. M., Industry and the Environment, Madsen, Essentials of Business Ethics

Klebe Trevino Linda, Butterfield Kenneth D., and McCabe Donald L., The Ethical Context in Organizations: Influences on Employee Attitudes and Behaviors, Business Ethics Quarterly, 1998 (3)

Klein E. R., The One Necessary Condition for a Successful Business Ethics Course: The Teacher Must Be a Philosopher, Business Ethics Quarterly, 1998 (3)

Klein Sherwin, Don Quixote and the Problem of Idealism and Realism in Business Ethics, Business Ethics Quarterly, 1998 (1)

Koehn Daryl, Virtue Ethics, the Firm, and Moral Psychology, Business Ethics Quarterly, 1998 (3)

Koslowski Peter F., The ethics of capitalism, Harvey, Business Ethics: a European Approach

Kuhlmann Eberhard, Customers, Harvey, Business Ethics: a European Approach

Kujala Johanna, Analysing Moral Issues in Stakeholder Relations – A Questionnaire Development Process, Business and Leadership Ethics, June 1998

Kupfer Andrew, Is Drug Testing Good or Bad?, Madsen, Essentials of Business Ethics

Kwame Safro, Doin' Business in an African Country, Journal of Business Ethics, Rae, Beyond Integrity

Lacour Jean-Philippe, Ces droles de tribunaux de commerce, Those funny courts called tribunaux de commerce, in French, La Tribune, 20 octobre 1999

Lambert Agnes, Les fonds ethiques s'ouvrent aux particuliers, Ehical funds open to the public, in French, La Tribune, 24.9.99

Laurent Philippe, Ethics, Money and Globalisation, Richardson, World Ethics Report on Finance and Money

Lea David, The Infelicities of Business Ethics in the Third World: The Melanesian Context, Business Ethics Quarterly, 1999 (3)

Lei Kai, New Banking Law in China, Richardson, World Ethics Report on Finance and Money

Leiser Burton M., Ethics and Equity in the Securities Industry, Williams, Ethics and the Investment Industry

Leithart Peter J., Snakes in the Garden: Sanctuaries, Sanctuary Pollution, and the Global Environment, Stewardship Journal, Rae, Beyond Integrity

Le Lien Charles, The Labours of Sisyphus – Going Beyond the Project for a Single Currency, Richardson, World Ethics Report on Finance and Money

Leroy Pierre-Henri, Shareholding and Society, Richardson, World Ethics Report on Finance and Money

Levitt Theodore, The Morality (?) of Advertising, Harvard Business Review, Rae, Beyond Integrity

Luijk van Henk, Business ethics: the field and its importance, Harvey, Business Ethics: a European Approach

Mackenzie Craig and Lewis Alan, Morals and Markets: The Case of Ethical Investing, Business Ethics Quarterly, 1999 (3)

Magnet Myron, The Decline and Fall of Business Ethics, Madsen, Essentials of Business Ethics

Mahoney Jack, How to be ethical: ethics resource management, Harvey, Business Ethics: a European Approach

Maitland Ian, Community Lost? Business Ethics Quarterly, 1998 (4)

Maitland Ian, The Limits of Business Self-Regulation, California Management Review, Rae, Beyond Integrity

256

Marens Richard and Wicks Andrew, Getting Real: Stakeholder Theory, Managerial Practice, and the General Irrelevance of Fiduciary Duties Owed to Shareholders, Business Ethics Quarterly, 1999 (2)

Margolis Joshua D., Psychological Pragmatism and the Imperative of Aims: A New Approach for Business Ethics, Business Ethics Quarterly, 1998 (3)

Marturano Marco, Italian Citizens' Confidence in the Judiciary and State Institutions, Richardson, World Ethics Report on Finance and Money

Marx Gary T., The Case of the Omniscient Organization, Ethics at Work, A Harvard Business Review Paperback, 1991

Mathiesen Kay, Game Theory in Business Ethics: Bad Ideology or Bad Press?, Business Ethics Quarterly, 1999 (1)

McCann Dennis P., "Accursed Internationalism" of Finance: Coping with the Resource of Catholic Social Teaching, Williams, Ethics and the Investment Industry

McClennen Edward F., Moral Rules as Public Goods, Business Ethics Quarterly, 1999 (1)

McCoy Bowen H., The Parable of the Sadhu, Madsen, Essentials of Business Ethics

McMahon Thomas F., Transforming Justice: A Conceptualization, Business Ethics Quarterly, 1999 (4)

Messick David M., Social Categories and Business Ethics, Business Ethics Quarterly, Special Issue #1, 1998

Michelman James H., Some Ethical Consequences of Economic Competition, Journal of Business Ethics, Rae, Beyond Integrity

Milgram Stanley, The Perils of Obedience, Obedience to Authority, Rae, Beyond Integrity

Missir di Lusignano Alessandro, Protecting the Financial Interests of the European Community and Fighting Financial Crime, Richardson, World Ethics Report on Finance and Money

Moberg Dennis J., The Big Five and Organizational Virtue, Business Ethics Quarterly, 1999 (2)

Morley Alfred C., Nurturing Professional Standards in the Investment Industry, Williams, Ethics and the Investment Industry

Morris Christofer W., What is This Thing Called "Reputation"?, Business Ethics Quarterly, 1999 (1)

Movahedi Nahid, Changes in Japanese Capitalism, Richardson, World Ethics Report on Finance and Money

Murphy Patrick E., Creating and Encouraging Ethical Corporate Structures, Sloan Management Review, Rae, Beyond Integrity

Nader Ralph, The Anatomy of Whistle-Blowing, Madsen, Essentials of Business Ethics

Nagel Thomas, A Defense of Affirmative Action, Senate Judiciary Committee, Rae, Beyond Integrity

Nash Laura L., Ethics without the sermon, Madsen, Essentials of Business Ethics

Nesteruk Jeffrey, Reimagining the Law, Business Ethics Quarterly, 1999 (4)

Neuville Colette, Protection judiciaire des actionnaires minoritaires, Legal protection of minority shareholders, in French, Ecole Nationale de la Magistrature, 12 mai 1997

Newton Lisa, The Hostile Takeover: An Opposition View, Ethical Theory and Business, Rae, Beyond Integrity

Nielsen Richard P., Can Ethical Character be Stimulated and Enabled? An Action Learning Approach to Teaching and Learning Organization Ethics, Business Ethics Quarterly, 1998 (3)

Novak Michael, A Theology of the Corporation, The Corporation, Rae, Beyond Integrity

Novak Michael, Virtuous Self-Interest, The Spirit of Democratic Capitalism, Rae, Beyond Integrity

O'Hara Patricia and Blakey Robert G., Legal Aspects of Insider Trading, Williams, Ethics and the Investment Industry

Olasky Marvin, Compassion, Religion and Liberty, Rae, Beyond Integrity

O'Neill June, An Argument Against Comparable Worth, Comparable Worth: An Issue for the 80's, Rae, Beyond Integrity

Orlando John, The Fourth Wave: The Ethics of Corporate Downsizing, Business Ethics Quarterly, 1999 (2)

Pastin Mark and Hooker Michael, Ethics and the Foreign Corrupt Practices Act, Madsen, Essentials of Business Ethics

Pastre Olivier, The Ten Commandments of Corporate Governance, Richardson, World Ethics Report on Finance and Money

Pava Moses L., Developing a Religiously Grounded Business Ethics: A Jewish Perspective, Business Ethics Quarterly, 1998 (1)

Perquel Jean-Jacques, New Markets, Richardson, World Ethics Report on Finance and Money

Pezard Alice, The Vienot Report on Corporate Governance, Richardson, World Ethics Report on Finance and Money

Pezard Alice, Confidence in the Judiciary, Richardson, World Ethics Report on Finance and Money

Phelan John J. Jr., Ethical Leadership and the Investment Industry, Williams, Ethics and the Investment Industry

Philips Robert A. and Margolis Joshua D., Toward an Ethics of Organizations, Business Ethics Quarterly, 1999 (4)

Pierenkemper Toni, The German Fear of Inflation, or Can History Teach Us Lessons?, Richardson, World Ethics Report on Finance and Money

Ploix Helene, Ethics and Financial Markets, Richardson, World Ethics Report on Finance and Money

Purdy Laura M., In Defense of Hiring Apparently Less Qualified Women, Journal of Social Philosophy, Rae, Beyond Integrity

Rak Pavle, Crime and Finance in Russia, Richardson, World Ethics Report on Finance and Money

Reed Darryl, Stakeholder Management Theory: A Critical Theory Perspective, Business Ethics Quarterly, 1999 (3)

Renard Vincent, Corruption and Real Estate in Japan, Richardson, World Ethics Report on Finance and Money

Rivoli Pietra, Ethical Aspects of Investor Behavior, Journal of Business Ethics, Rae, Beyond Integrity

Robin Donald, Giallourakis Michael, David Fred R., and Moritz Thomas, A Different Look at Codes of Ethics, Madsen, Essentials of Business Ethics

Roma Giuseppe, Italy's Moneylenders, Between Illegality and Social Compromise, Richardson, World Ethics Report on Finance and Money

Rorty Richard, Can American Egalitarianism Survive a Globalized Economy?, Business Ethics Quarterly, Special Issue #1, 1998

Russell James W., A Borderline Case: Sweatshops Cross the Rio Grande, Madsen, Essentials of Business Ethics

Sass Steven, Risk at the PBGC, Richardson, World Ethics Report on Finance and Money

Schermerhorn Jr. John R., Terms of Global Business Engagement in Ethically Challenging Environments: Applications to Burma, Business Ethics Quarterly, 1999 (3)

Schneider Jacques-Andre, Pension Fund Management and the Ethics of Responsibility, Richardson, World Ethics Report on Finance and Money

Schokkaert Erik and Eyckmans Johan, Environment, Harvey, Business Ethics: a European Approach

Schumacher E. F., Buddhist Economics, Small Is Beautiful, Rae, Beyond Integrity

Sciarelli Sergio, Corporate Ethics and the Entrepreneurial Theory of "Social Success", Business Ethics Quarterly, 1999 (4)

Senate Finance Commission, Stock Options in France, Richardson, World Ethics Report on Finance and Money

Servet Jean-Michel, Metamorphosis of a Chinese Dollar, Richardson, World Ethics Report on Finance and Money

Sethi Prakash S. and Sama Linda M., Ethical Behavior as a Strategic Choice by Large Corporations: The Interactive Effect of Marketplace Competition, Industry Structure and Firm Resources, Business Ethics Quarterly, 1998 (1)

Seymour Sally, The Case of the Willful Whistle-Blower, Ethics at Work, A Harvard Business Review Paperback, 1991

Seymour Sally, The Case of the Mismanaged Ms., Ethics at Work, A Harvard Business Review Paperback, 1991

Sharp Paine Lynn, Managing for Organizational Integrity, Harvard Business Review, Rae, Beyond Integrity

Shaw Bill, Aristotle and Posner on Corrective Justice: The Tortoise and the Hare, Business Ethics Quarterly, 1999 (4)

Shaw Bill, Community: A Work in Progress, Business Ethics Quarterly, 1998 (4)

Shaw Bill, Should Insider Trading Be Outside the Law?, Business and Society Review, Rae, Beyond Integrity

Shriver Donald W. Jr., Ethical Discipline and Religious Hope in the Investment Industry, Williams, Ethics and the Investment Industry

Singer M. S., Paradigms Linked: A Normative-Empirical Dialogue about Business Ethics, Business Ethics Quarterly, 1998 (3)

Sirico Robert Fr., The Entrepreneurial Vocation, Acton Institute, Rae, Beyond Integrity

Skillen James W., Common Moral Ground and the Natural Law Argument, Rae, Beyond Integrity

Smith H. R. and Carroll Archie B., Organizational Ethics: A Stacked Deck, Journal of Business Ethics, Rae, Beyond Integrity

Smith Virgil, The Place of Character in Corporate Structure, Rae, Beyond Integrity

Smith William, A View from Wall Street, Williams, Ethics and the Investment Industry

Snell Robin S., Obedience to Authority and Ethical Dilemmas in Hong Kong Companies, Business Ethics Quarterly, 1999 (3)

Solomon Robert C., Game Theory as a Model for Business and Business Ethics, Business Ethics Quarterly, 1999 (1)

Solomon Robert C., The Moral Psychology of Business: Care and Compassion in the Corporation, Business Ethics Quarterly, 1998 (3)

Solomon Robert C., Business with Virtue: Maybe Next Year?, Business Ethics Quarterly, 2000 (1)

Soule Edward, Trust and Managerial Responsibility, Business Ethics Quarterly, 1998 (2)

Steele Shelby, Affirmative Action: The Price of Preference, The Content of Our Character, Rae, Beyond Integrity

Takala Tuomo, Postmodern Challenge to Business Ethics, Business and Leadership Ethics, June 1998

Thiery Nicolas, A la decouverte des fonds ethiques, The discovery of ethical funds, in French, La Tribune, 19.10.99

Thiveaud Jean-Marie, Confidence Reigns Supreme, Richardson, World Ethics Report on Finance and Money

Tierney Paul E. Jr., The Ethos of Wall Street, Williams, Ethics and the Investment Industry

Trichet Jean-Claude, Is There an Increase in Risks to the System and How Should We Confront It, Richardson, World Ethics Report on Finance and Money

Tumminen Rauno, Ownership in Environmental Management, Business and Leadership Ethics, June 1998

Uusitalo Eeva and Outi, Marketing Ethics, Business and Leadership Ethics, June 1998

Vanderschraaf Peter, Hume's Game-Theoretic Business Ethics, Business Ethics Quarterly, 1999 (1)

Vanderschraaf Peter, Introduction: Game Theory and Business Ethics, Business Ethics Quarterly, 1999 (1)

Vandivier Kermit, Whu Should My Conscience Bother Me?, In the Name of Profit, Rae, Beyond Integrity

Velasquez Manuel G., Corporate Ethics: Losing It, Having It, Getting It, Madsen, Essentials of Business Ethics

Velasquez Manuel, Globalization and the Failure of Ethics, Business Ethics Quarterly, 2000 (1)

Vidaver-Cohen Deborah, Moral Imagination in Organizational Problem-Solving: An Institutional Perspective, Business Ethics Quarterly, Special Issue #1, 1998

Vidaver-Cohen Deborah, Motivational Appeal in Normative Theories of Enterprise, Business Ethics Quarterly, 1998 (3)

Virard Marie-Paule, Companies: the Hidden Side of the Accounts, Richardson, World Ethics Report on Finance and Money

Waide John, The Making of Self and World in Advertising, Journal of Business Ethics, Rae, Beyond Integrity

Wallis Jim, The Powerful and the Powerless, Agenda for Biblical People, Rae, Beyond Integrity

Walton Clarence C., Investment Bankers from Ethical Perspectives... With Special Emphasis on the Theory of Agency, Williams, Ethics and the Investment Industry

Warner Alison, Banks in a Spin, Richardson, World Ethics Report on Finance and Money

Warsh David, How Selfish Are People - Really?, Ethics at Work, A Harvard Business Review Paperback, 1991

Watson George W., Shefard Jon M., Stephens Carroll U., and Christman John C., Ideology and the Economic Social Contract in a Downsizing Environment, Business Ethics Quarterly, 1999 (4)

Watson Jr. Thomas S., Connecting People: Alternative Futures, Business and Leadership Ethics, June 1998

Weaver Gary R. and Klebe Trevino Linda, Compliance and Values Oriented Ethics Programs: Influences on Employees' Attitudes and Behavior, Business Ethics Quarterly, 1999 (2)

Weithers John G., Ethics within the Securities Industry, Williams, Ethics and the Investment Industry

Wensveen Siker Louke van, Christ and Business, Journal of Business Ethics, Rae, Beyond Integrity

Werhane Patricia H., A Bill of Rights for Employees and Employers, Madsen, Essentials of Business Ethics

Werhane Patricia H., Employee and Employer Rights in an Institutional Context, Ethical Theory in Business, Rae, Beyond Integrity

Werhane Patricia H., Moral Imagination and the Search for Ethical Decision Making in Management, Business Ethics Quarterly, Special Issue #1, 1998

Werhane Patricia H., The Ethics of Insider Trading, Journal of Business Ethics, Rae, Beyond Integrity

Werhane Patricia H., Exporting Mental Models: Global Capitalism in the 21st Century, Business Ethics Quarterly, 2000 (1)

Wicks Andrew, How Kantian a Kantian Theory of Capitalism?, Business Ethics Quarterly, Special Issue #1, 1998

Wicks Andrew C. and Glezen Paul L., In Search of Experts: A Conception of Expertise for Business Ethics Consultation, Business Ethics Quarterly, 1998 (1)

Wilmouth Robert K., Futures Market and Self-Regulation, Williams, Ethics and the Investment Industry

Wokutch Richard E. and Shepard Jon M., The Maturing of the Japanese Economy: Corporate Social Responsibility Implications, Business Ethics Quarterly, 1999 (3)

Wood Donna J., Ingroups and Outgroups: What Psychology Doesn't Say, Business Ethics Quarterly, Special Issue #1, 1998

Wu Xinwen, Business Ethical Perceptions of Business People in East China: An Empirical Study, Business Ethics Quarterly, 1999 (3)

INDEX